MONOGRAPHS ON THE PHYSICS AND CHEMISTRY OF MATERIALS

General Editors
H. FRÖHLICH, A. J. HEEGER,
P. B. HIRSCH, N. F. MOTT
R. BROOK

MONOGRAPHS ON THE
PHYSICS AND CHEMISTRY OF MATERIALS

Neutron diffraction (3rd edn) G. E. Bacon
Strong solids (3rd edn) A. Kelly and N. H. Macmillan
Nonequilibrium thermodynamics and its statistical foundations
  H. J. Kreuzer
Chemistry of the metal–gas interface M. W. Roberts and C. S. McKee
Quantum theory of collective phenomena G. L. Sewell
Experimental high-resolution electron microscopy J. C. H. Spence
Experimental techniques in low-temperature physics Guy K. White
Principles of dielectrics B. K. P. Scaife
Optical spectroscopy of inorganic solids B. Henderson and G. F. Imbusch

THEORY OF DIELECTRICS

DIELECTRIC CONSTANT
AND
DIELECTRIC LOSS

BY

H. FRÖHLICH

PROFESSOR OF THEORETICAL PHYSICS
IN THE UNIVERSITY OF LIVERPOOL

SECOND EDITION

OXFORD
AT THE CLARENDON PRESS

Oxford University Press, Walton Street, Oxford OX2 6DP
Oxford New York Toronto
Delhi Bombay Calcutta Madras Karachi
Petaling Jaya Singapore Hong Kong Tokyo
Nairobi Dar es Salaam Cape Town
Melbourne Auckland
and associated companies in
Berlin Ibadan

Oxford is a trade mark of Oxford University Press

Published in the United States
by Oxford University Press, New York

© Oxford University Press, 1958

First published 1949
Reprinted lithographically in Great Britain
at the University Press, Oxford, 1950
from corrected sheets of the first edition
Second edition 1958
Reprinted 1963, 1968
First published in paperback 1986
Reprinted 1990

All rights reserved. No part of this publication may be reproduced,
stored in a retrieval system, or transmitted, in any form or by any means,
electronic, mechanical, photocopying, recording, or otherwise, without
the prior permission of Oxford University Press

This book is sold subject to the condition that it shall not, by way
of trade or otherwise, be lent, re-sold, hired out or otherwise circulated
without the publisher's prior consent in any form of binding or cover
other than that in which it is published and without a similar condition
including this condition being imposed on the subsequent purchaser

British Library Cataloguing in Publication Data
Fröhlich, H.
Theory of dielectrics: dielectric constant
and dielectric loss.—2nd ed.—
(Monographs on the physics and chemistry of materials)
1. Dielectrics
I. Title II. Series
537'.24 QC585
ISBN 0–19–851379–8

Printed and bound in Great Britain by
Biddles Ltd, Guildford and King's Lynn

PREFACE TO FIRST EDITION

PROPERTIES of dielectric materials are of interest to scientists in various branches: physicists, chemists, electrical engineers, biologists. Their interests concern different aspects; thus an electrical engineer will require the dependence of dielectric loss on frequency and temperature in order to find a substance which is nearly loss free in a certain range. The chemist on the other hand can use this knowledge to draw conclusions on the properties of molecules. For these, and for many other purposes, it is imperative to have a theory of dielectrics.

The present book is intended to give a systematic account of the theory of the dielectric constant and of dielectric loss. This it is hoped will satisfy the requirements of the various branches of research interested in dielectrics. In writing this account I found that the subject deserves interest also from a methodical point of view as an application of classical statistical mechanics. That this application is far from trivial is shown by some of the controversies in the literature which have lasted until very recent years; also the general theorems derived in § 7 seem to be novel.

It was my intention to write this monograph for the use of applied scientists. I hope, however, that the sections dealing with the general theory will also be of value to students. The required mathematical technique only occasionally exceeds acquaintance with calculus; even so I have been told that its extensive use might be too heavy for biologists. The reader is assumed to have a certain elementary knowledge of atomic and molecular physics, statistical mechanics, and electrostatics. Quantum mechanics will not be required; its relation to the theory of dielectrics is discussed in van Vleck's book ($V1$).

Units unless stated otherwise refer to the electrostatic c.g.s. system. Vectors are represented by bold type. Unfortunately it was not always possible to avoid repetition of symbols. The meaning of the symbols $\propto$, $\simeq$, $\sim$ is respectively proportional to, approximately equal to, order of magnitude of. As usual, k is Boltzmann's constant, h is Planck's constant, $\hbar = h/2\pi$.

I am grateful to a great number of colleagues for help of various kinds. To Dr. F. C. Frank, Professor Willis Jackson, Dr. H. Pelzer for reading all, or parts of, the manuscript, and making valuable suggestions; to Dr. R. Sack for his help in reading the proofs; to Mr. S. Zienau for making the index. My particular thanks, however, are due to Dr. B. Szigeti who helped to collect the experimental material, and to Dr. J. H. Simpson who read the first draft of the manuscript and made many useful suggestions.

I should like to use this opportunity to express my appreciation to the British Electrical and Allied Industries Research Association (E.R.A.) without whose support much of the work described here would have remained undone.

I also wish to express my thanks to the Physical Society, the Faraday Society, *Nature*, and to the authors quoted for the use of illustrations.

H. F.

PREFACE TO THE SECOND EDITION

This second edition differs from the first one through an addendum B containing three paragraphs. The first two of these deal with the general theory of the static dielectric constant. They develop the general theory farther than has been done in § 7. In the last paragraph, the important work of R. A. Sack and E. P. Gross on dielectric loss is mentioned, though at present this has not yet reached a stage for detailed treatment in a relatively elementary book.

Liverpool
1957

H. F.

CONTENTS

CHAPTER I. MACROSCOPIC THEORY	1
§ 1. Static fields	1
§ 2. Time-dependent fields	4
§ 3. Energy and entropy	9
CHAPTER II. STATIC DIELECTRIC CONSTANT	15
§ 4. Survey	15
§ 5. Dipolar interaction	21
§ 6. Dipolar molecules in gases and dilute solutions	26
§ 7. General theorems	36
§ 8. Special cases	48
CHAPTER III. DYNAMIC PROPERTIES	62
§ 9. The establishment of equilibrium	62
§ 10. The Debye equations	70
§ 11. Models for the Debye equations	78
§ 12. Generalizations	90
§ 13. Resonance absorption	98
CHAPTER IV. APPLICATIONS	104
§ 14. Structure and dielectric properties	104
§ 15. Non-polar substances	109
§ 16. Dipolar substances; gases and dilute solutions	115
§ 17. Dipolar solids and liquids	130
§ 18. Ionic crystals	149
APPENDIX	160
A 1. Electro-magnetic theory	160
A 2. Dipole moments and other electrostatic problems	163
A 3. The Clausius-Mossotti formula	169
A 4. Shape of absorption lines	173
B 1. Static dielectric constant	176
B 2. Reaction fields: a simple example	183
B 3. Dielectric loss	187
REFERENCES	188
INDEX	191

CHAPTER I

MACROSCOPIC THEORY

1. Static fields

THE electric properties of dielectric substances are usually described in terms of the dielectric constant. For most materials this quantity is independent of the strength of the electric field over a wide range of the latter, but in the case of alternating fields depends on the frequency. It also depends on parameters, such as the temperature, which define the state of the material. In the macroscopic (phenomenological) theory, which will be summarized in the present chapter, the dielectric constant is supposed to be known empirically. The purpose of the rest of the book will be to derive the dielectric constant (and its variation with temperature, frequency, etc.) from the atomic structure of the material.

Throughout this book we shall be interested in homogeneous substances only, and in electric fields which are independent of the space coordinates, although they may depend on time.

Consider now a condenser consisting of two parallel plates in vacuum whose distance apart d is small compared with their linear dimensions, and suppose that they are charged electrically with charges $+\sigma A$ and $-\sigma A$ respectively. A is the surface area of a plate, and hence $+\sigma$ is the charge per unit area which will be denoted as surface density. The charges give rise to an electric field which inside the condenser is practically homogeneous and directed perpendicular to the surface. Its amount is given by
$$E = 4\pi\sigma, \text{ (vacuum)}. \tag{1.1}$$

In this equation the factor 4π is due to the particular way in which the unit of electric charge is defined, leading to the electrostatic c.g.s. system of units usually chosen in atomic physics. Equation 1.1 should be completed by the definition of the electric field as the force (in dynes) acting on the unit of electric charge which then leads to a unique definition of this unit.

Between the two plates of the condenser a voltage can be measured whose absolute value Φ is given by

$$\Phi = |E|d. \tag{1.2}$$

Now suppose the space between the plates to be filled with a homogeneous dielectric material while the charge on the plates remains unaltered. This will cause the voltage to drop to a smaller value, and the ratio of its former value to the present one is denoted by the static dielectric constant ϵ_s. Since equation 1.2 still holds, the electric field strength has also decreased, its present value being

$$E = 4\pi\sigma/\epsilon_s. \tag{1.3}$$

From equations 1.1 and 1.3 it follows that the drop in the field strength accompanying the insertion of the dielectric might also be achieved by reducing the surface-density σ of the electric charge by the amount

$$P = \sigma\left(1 - \frac{1}{\epsilon_s}\right) = \sigma\frac{\epsilon_s - 1}{\epsilon_s}. \tag{1.4}$$

Therefore the influence of the electric field on the dielectric is equivalent to charging the two surfaces of the dielectric with charges of opposite sign in such a way that the positive condenser plate is faced by the negatively charged surface of the dielectric, and vice versa. The surface density of the charge is constant, and amounts to P. This behaviour of the dielectric is to be expected from the atomic point of view according to which any substance containing no net charge consists of an equal number of positive and negative elementary charges. In a dielectric in particular these charges cannot move freely through the medium (as in a conductor), but they can be displaced. Clearly negative charges will be displaced towards the positive plate and conversely. The total charge passing through any unit of area within the dielectric, which is parallel to the condenser plates, is the same and its amount is equal to P. P, therefore, is called the polarization of the dielectric.

In macroscopic physics the introduction of the polarization P through the displacement of charges is rather fictitious because these charges cannot be removed from the dielectric. They compensate charges of opposite sign but equal absolute value of the condenser plates. These latter charges are sometimes denoted as 'bound' charges, whereas the original charges at the condenser plates are called 'true' charges (cf. Fig. 1). It is customary then to introduce a new field-quantity D, which is described in terms of the true charge. This is the electric displacement, defined by

$$D = 4\pi\sigma. \tag{1.5}$$

In vacuum, therefore, $D = E$. In dielectrics, however, using 1.3 and 1.5

$$D = \epsilon_s E. \tag{1.6}$$

Clearly from equations 1.4, 1.5, and 1.6

$$D = E + 4\pi P. \tag{1.7}$$

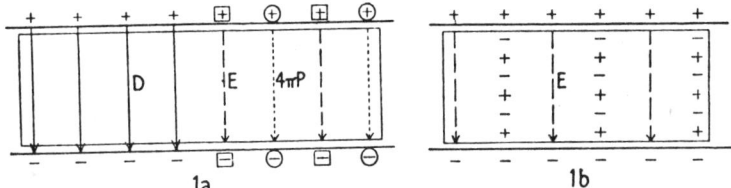

Fig. 1. Dielectric material with dielectric constant $\epsilon_s = 2$ between condenser plates. (a) *Macroscopic description.* The left-hand side of the figure shows true charges $+$, $-$ as sources of D. The right-hand side shows that the true charges can be considered as composed of bound charges $\oplus$, $\ominus$ and of free charges $\boxed{+}$, $\boxed{-}$ being sources of $4\pi P$ and E respectively, and leading to

$$D = E + 4\pi P.$$

(b) *Atomic description.* Only true charges exist, and the field is described by E only; the polarization of the dielectric (indicated by $+ - + ... -$), however, leads to surface charges which compensate some of the charges on the condenser plates.

Thus in macroscopic physics the electric field in a dielectric must be described by two field quantities. The electric field-strength E and the electric displacement D are usually chosen, and the polarization P can then be derived with the help of equation 1.7. D is defined by the (true) charge according to equation 1.5, and E can be derived from D with the help of a special relation 1.6 which is characteristic for the particular dielectric material.

To link up this description with atomic physics it should be noted that the surface charges $\pm PA$ give rise to an electric dipole moment M of the dielectric given by

$$M = PAd = PV, \qquad (1.8)$$

where $V = Ad$ is its volume. On the other hand, as shown in § 4, this electric moment M can be calculated from the configuration of the positive and negative elementary charges which form the constituents of the substance. Thus from 1.8 with the use of equations 1.6 and 1.7 it follows that

$$\epsilon_s - 1 = 4\pi M/VE = 4\pi P/E. \qquad (1.9)$$

This equation provides the required link between macroscopic and atomic theory.

2. Time-dependent fields

Now consider that the charges on the condenser plates, and hence the electric field, depend on time. As in the static case, a dielectric which is placed between the plates will be polarized by the field. The displacement of charges connected with this polarization usually shows some inertia. Thus if a constant field is suddenly applied the polarization will not reach its static value immediately, but will approach it gradually (cf. Fig. 2).

As in the static case, two field quantities are required to describe the electric field inside the dielectric and one usually chooses the electric field-strength E and the electric displacement D. The latter is still defined by equation 1.5, and the correlation between E, D, and P given by equation 1.7 still holds. Equation 1.6, however, correlating E and D, is no longer valid in the present case, but has to be replaced by a more general relation.

Consider first the important case of a periodic field, e.g.

$$E = E_0 \cos \omega t, \qquad (2.1)$$

where E_0 is independent of time and $\omega/2\pi$ is the frequency in cycles per second. If a field of this type has persisted for a sufficient length of time, D too must be periodic in time. In general, however, D will not necessarily be in phase with E, but will show a phase-shift ϕ, i.e.

$$D = D_0 \cos(\omega t - \phi) = D_1 \cos \omega t + D_2 \sin \omega t, \qquad (2.2)$$

where according to elementary trigonometry

$$D_1 = D_0 \cos\phi, \qquad D_2 = D_0 \sin\phi. \tag{2.3}$$

For most dielectrics D_0 will be proportional to E_0, but the ratio D_0/E_0 usually depends on frequency. Therefore, two

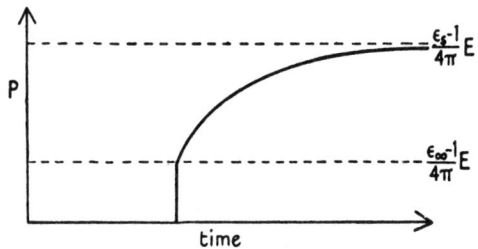

Fig. 2. Time-dependence of the polarization P of a dielectric when a constant electric field E is suddenly applied to it.

different dielectric constants, $\epsilon_1(\omega)$ and $\epsilon_2(\omega)$, both frequency dependent, can be introduced by

$$D_1 = \epsilon_1 E_0 \quad \text{and} \quad D_2 = \epsilon_2 E_0. \tag{2.4}$$

Thus from 2.3 and 2.4

$$\tan\phi = \frac{\epsilon_2}{\epsilon_1}. \tag{2.5}$$

It will be shown in § 3 that ϵ_2 is proportional to the energy loss in dielectrics.

As the frequency approaches zero the present description must become identical with that given in § 1. Thus (assuming that there exists no dielectric loss in static fields)

$$\epsilon_2(\omega) \to 0, \qquad \epsilon_1(\omega) \to \epsilon_s, \qquad \text{as} \quad \omega \to 0. \tag{2.6}$$

We shall add the further relation (cf. also § 10)

$$\epsilon_1(\omega) \to \epsilon_\infty, \qquad \text{as} \quad \omega \to \infty, \tag{2.7}$$

which is to be understood in such a way that ϵ_∞ is the value which $\epsilon_1(\omega)$ approaches at the highest frequencies contemplated in the present book. They correspond to wave-lengths in the infra-red region.

The above equations can be written in a condensed form by introducing a complex dielectric constant

$$\epsilon = \epsilon_1 + i\epsilon_2 \tag{2.8}$$

and replacing equation 2.1 by

$$E = E_0 e^{-i\omega t}, \tag{2.9}$$

considering, however, only the real part of this equation (which is identical with 2.1). Then the real part of the equation

$$D = \epsilon E \tag{2.10}$$

is identical with equations 2.2 and 2.4.

The two dielectric constants ϵ_1 and ϵ_2 if considered as functions of the frequency ω are not entirely independent if the relationship between E and D is a linear one.† This linear relationship is usually expressed by the principle of superposition and is best explained with the help of a more general time dependence of the field than has been considered above. Assume that during the time interval between u and $u+du$ an electric field of strength $E(u)$ has been applied to the dielectric and that the electric field vanishes outside this time interval. A displacement D will result which in view of the inertia of the polarization P will persist at times $t > u+du$, but which will gradually vanish.‡ D is thus a function of $t-u$, i.e.

$$D(t-u) = E(u)\alpha(t-u)\,du \quad \text{if} \quad t > u+du,$$

where $\alpha(t-u)$ is the decay function describing the gradual decrease of D, in particular

$$\alpha(t-u) \to 0 \quad \text{if} \quad t \to \infty. \tag{2.11}$$

The displacement D contains a part which can follow the field practically immediately, and which in view of the meaning given to ϵ_∞ will be assumed to be equal to $\epsilon_\infty E(u)$.

Thus

$$D(t-u) = \epsilon_\infty E(u) + E(u)\alpha(0)\,du \quad \text{if} \quad u < t < u+du,$$

† Cf. B. Gross [G4] and S. Whitehead [W5], where further references are given.
‡ Note that according to 1.7 $D = 4\pi P$ if $E = 0$.

where α may be considered to remain at the value $\alpha(0)$ during the short interval du.

Suppose now that at a later time interval between u' and $u'+du'$ another field $E(u')$ is applied. Then by the principle of superposition it will be assumed that the corresponding displacement $D(t-u')$ is superposed linearly on the former one. This principle of superposition if applied to a continuous time-dependent field $E(u)$ initiated at the time $u = 0$ requires that the displacement $D(t)$ at the time t is given by

$$D(t) = \epsilon_\infty E(t) + \int_0^t E(u)\alpha(t-u)\,du. \tag{2.12}$$

This equation will now be applied to periodic fields. Introducing E from equation 2.1 into 2.12 thus leads to

$$D(t) - \epsilon_\infty E_0 \cos\omega t = E_0 \int_0^t \alpha(t-u)\cos\omega u\,du$$

$$= E_0 \int_0^t \alpha(x)\cos\omega(t-x)\,dx$$

if $x = t-u$ is introduced. It should be noted that in all integrations t is to be considered as a parameter. Again it will be assumed that the field has persisted sufficiently long to make D a periodic function of time. This means that t is larger than the time t_0 at which $\alpha(t)$ practically vanishes. Then in view of 2.11 the above integration over the coordinate x can be extended to infinity without appreciably altering the value of the integral, i.e.

$$D(t) - \epsilon_\infty E_0 \cos\omega t = E_0 \int_0^\infty \alpha(x)\cos\omega(t-x)\,dx,$$

or applying a simple trigonometric formula,

$$D(t) - \epsilon_\infty E_0 \cos\omega t = E_0 \cos\omega t \int_0^\infty \alpha(x)\cos\omega x\,dx +$$

$$+ E_0 \sin\omega t \int_0^\infty \alpha(x)\sin\omega x\,dx. \tag{2.13}$$

Comparing this expression with equations 2.2 and 2.4 it follows that

$$\epsilon_1(\omega) - \epsilon_\infty = \int_0^\infty \alpha(x)\cos \omega x \, dx \qquad (2.14)$$

$$\epsilon_2(\omega) = \int_0^\infty \alpha(x)\sin \omega x \, dx. \qquad (2.15)$$

Both functions $\epsilon_1(\omega) - \epsilon_\infty$ and $\epsilon_2(\omega)$ can thus be derived from the same function $\alpha(x)$ and, therefore, cannot be independent.

A calculation carried out in the appendix (A.1.iii) shows that

$$\epsilon_1(\omega) - \epsilon_\infty = \frac{2}{\pi} \int_0^\infty \epsilon_2(\mu) \frac{\mu}{\mu^2 - \omega^2} \, d\mu, \qquad (2.16)$$

and
$$\epsilon_2(\omega) = \frac{2}{\pi} \int_0^\infty \{\epsilon_1(\mu) - \epsilon_\infty\} \frac{\omega}{\omega^2 - \mu^2} \, d\mu, \qquad (2.17)$$

where μ is a variable of integration. Both integrals are principal values.

Equation 2.16 can be used to calculate the static dielectric constant from $\epsilon_2(\omega)$. In this case clearly

$$\epsilon_s = \epsilon_1(0) = \epsilon_\infty + \frac{2}{\pi} \int_0^\infty \epsilon_2(\mu) \frac{d\mu}{\mu}, \qquad (2.18)$$

which indicates that substances for which $\epsilon_s - \epsilon_\infty$ is very small cannot show appreciable dielectric losses (which are proportional to ϵ_2).

Finally, it should be mentioned that the correlation between macroscopic and atomic theory is provided by an equation similar in nature to equation 1.9. It is based on the fact that equation 1.7 holds in the time-dependent case as well as in the static one. With a similar argument to that in § 1 it follows that the polarization P is equal to the electric moment M per unit volume, leading to equation 1.8. Introducing the complex dielectric constant ϵ with the help of 2.10, using 1.7, it follows that
$$(\epsilon - 1)E = 4\pi M/V, \qquad (2.19)$$

in which only the real part of the expression on the left-hand side is considered. Thus if

$$M = M_1 \cos \omega t + M_2 \sin \omega t, \qquad (2.20)$$

then $\quad \epsilon_1 - 1 = 4\pi M_1/VE_0, \qquad \epsilon_2 = 4\pi M_2/VE_0. \qquad (2.21)$

3. Energy and Entropy

A. *Static fields*†

In many text-books the expression

$$\epsilon_s \frac{E^2}{8\pi}$$

is stated to represent the electric energy per unit volume of a dielectric material with a static dielectric constant ϵ_s in the presence of an electric field E. This statement is misleading whenever ϵ_s depends on temperature. In fact it suggests that the energy difference per unit volume of the dielectric, first in the presence and then in the absence of an electric field, is always given by the above expression. This energy difference ought, however, to depend on the state in which the dielectric is kept while the electric field is applied; this might, for instance, be done isothermally or adiabatically. A more accurate discussion given below shows that the above expression actually is the change of the free energy of the dielectric.

Before this discussion is commenced the reader will briefly be reminded of the two fundamental laws of thermodynamics. Consider as a simple example a gas of volume v, pressure p, and temperature T. By expanding it by the small volume dv, work amounting to pdv will be done. Conservation of energy, therefore, requires that

$$dU = dQ - pdv \qquad (3.1)$$

is the change in the energy content of the gas if dQ is the influx of heat during the expansion. Equation 3.1 represents the first law of thermodynamics in this simple case. An analysis shows that the quantity dQ is not a total differential, i.e. that no unique function Q of the variables exists such that dQ is the

† Cf. Abraham-Becker [*A1*, Chapter XI].

difference between two neighbouring values Q_1 and Q_2. For a reversible process, the expression

$$dS = \frac{dQ}{T} \tag{3.2}$$

does, however, represent a total differential of a function S, the entropy. S is of fundamental importance in connexion with the second law of thermodynamics (cf. text-books).

With the help of S the Helmholtz free energy F can be derived by
$$F = U - TS. \tag{3.3}$$
It represents the maximum amount of work which the system can be made to do in an isothermal (i.e. constant temperature) process.

In the case of dielectric material in an electric field, electromagnetic theory (cf. Appendix A 1.i) shows that the quantity

$$\frac{1}{4\pi} E\, dD \tag{3.4}$$

represents the influx of energy into the dielectric (per unit volume) if the displacement D is increased by the small amount dD.

Assume now that the volume of the dielectric is always kept constant, and that the temperature T is the only parameter besides the electric field E considered to be varied. Then the increase dU of the energy U per unit volume of the dielectric in a process in which either T or E, or both, are varied slightly is given by
$$dU = dQ + \frac{E}{4\pi} dD \tag{3.5}$$
if dQ is the influx of heat per unit volume.

This equation is similar in structure to equation 3.1 for a gas if E and D are replaced by $-p$ and v. The relation existing in gases between p, v, and T (the equation of state) is, however, different from the relation between E, D, and T. For the latter, equation 1.6 will be supposed to hold with a dielectric constant ϵ_s which may depend on T but is independent of E. Thus

$$dD = d(\epsilon_s E) = \epsilon_s\, dE + E\, d\epsilon_s = \epsilon_s\, dE + E \frac{\partial \epsilon_s}{\partial T} dT,$$

ENERGY AND ENTROPY

which means that a variation of D may be due to a change in the field-strength E at constant temperature, and to a change in temperature at constant E. For the following it will be useful to take T and E^2 as the independent variables. Equation 3.5 representing the first law of thermodynamics then becomes

$$dQ + \frac{\epsilon_s}{8\pi} d(E^2) + \frac{E^2}{4\pi} \frac{\partial \epsilon_s}{\partial T} dT = dU = \frac{\partial U}{\partial (E^2)} d(E^2) + \frac{\partial U}{\partial T} dT. \tag{3.6}$$

A further relation will now be obtained from the entropy law according to which dS, given by equation 3.2, is a total differential. This means that a unique function $S(T, E^2)$ must exist such that

$$dS = \frac{\partial S}{\partial T} dT + \frac{\partial S}{\partial (E^2)} d(E^2). \tag{3.7}$$

Thus if it is found that

$$dS = A(T, E^2) dT + B(T, E^2) d(E^2), \tag{3.8}$$

where A and B are both functions of the two variables T and E^2, the condition that dS is a total differential requires that

$$\frac{\partial B}{\partial T} = \frac{\partial A}{\partial (E^2)} \tag{3.9}$$

because both sides of the equation 3.9 are equal to $\partial^2 S/\partial T \partial(E^2)$. Now according to 3.2, inserting dQ from 3.6,

$$dS = \frac{1}{T}\left(\frac{\partial U}{\partial T} - \frac{E^2}{4\pi} \frac{\partial \epsilon_s}{\partial T}\right) dT + \frac{1}{T}\left(\frac{\partial U}{\partial (E^2)} - \frac{\epsilon_s}{8\pi}\right) d(E^2). \tag{3.10}$$

This equation is of the type 3.8, and equation 3.9, therefore, becomes

$$\frac{\partial}{\partial T}\left\{\frac{1}{T}\left(\frac{\partial U}{\partial (E^2)} - \frac{\epsilon_s}{8\pi}\right)\right\} = \frac{\partial}{\partial (E^2)}\left\{\frac{1}{T}\left(\frac{\partial U}{\partial T} - \frac{E^2}{4\pi} \frac{\partial \epsilon_s}{\partial T}\right)\right\}.$$

Carrying out the differentiations one finds

$$\frac{\partial U}{\partial (E^2)} = \frac{1}{8\pi}\left(\epsilon_s + T \frac{\partial \epsilon_s}{\partial T}\right).$$

Integrating with respect to E^2 yields the energy density

$$U = U_0(T) + \left(\epsilon_s + T \frac{\partial \epsilon_s}{\partial T}\right)\frac{E^2}{8\pi}, \tag{3.11}$$

where $U_0(T)$ is independent of E^2 but depends on T and thus represents the energy of the dielectric in the absence of a field.

The entropy S can now easily be calculated because on comparing equation 3.10 with 3.7, both $\partial S/\partial T$ and $\partial S/\partial(E^2)$ are known if U is introduced from equation 3.11. Thus

$$\frac{\partial S}{\partial T} = \frac{1}{T}\frac{\partial U_0}{\partial T} + \frac{E^2}{8\pi}\frac{\partial^2 \epsilon_s}{\partial T^2}, \qquad \frac{\partial S}{\partial(E^2)} = \frac{1}{8\pi}\frac{\partial \epsilon_s}{\partial T},$$

or integrating
$$S = S_0(T) + \frac{\partial \epsilon_s}{\partial T}\frac{E^2}{8\pi}, \tag{3.12}$$

where $S_0(T)$ is the entropy in the absence of a field. From equation 3.3 one finally finds for the free energy

$$F = F_0(T) + \frac{\epsilon_s E^2}{8\pi}, \tag{3.13}$$

where $F_0(T)$ is the free energy in the absence of a field. This proves our original contention.

The above expressions for U, S, and F are very instructive. Thus equation 3.13 for the free energy shows (in analogy to the meaning of F in gases) that the amount of electric energy available in an isothermal reversible process is $\epsilon_s E^2/8\pi$.

From equation 3.11 for the energy it follows that for substances such as dilute dipolar gases for which (ϵ_0 is independent of T)
$$\epsilon_s = \epsilon_0 + \text{constant}/T, \tag{3.14}$$

the energy change due to the field is given by $\epsilon_0 E^2/8\pi$. Thus in this case the temperature-dependent part of ϵ_s does not make any contribution to the energy. The remaining free energy $(\epsilon_s - \epsilon_0)E^2/8\pi$ available besides $\epsilon_0 E^2/8\pi$ is thus entirely due to a change in entropy.†

Finally, equation 3.12 shows that the entropy is increased by the field if $\partial \epsilon_s/\partial T$ is positive, and decreased if this quantity is negative. Since the entropy is a measure of the molecular disorder, an external field creates order in dipolar liquids and gases for which ϵ_s decreases with increasing T. This may be expected

† Near the absolute zero of temperature this would contradict the third law of thermodynamics; a temperature dependence of the type 3.14 can, therefore, not be valid near $T = 0$.

because the field will orientate some of the dipoles which in the absence of a field are at random. In some dipolar solids, on the other hand, ϵ_s increases with T, which means that an external field increases the disorder. This too is understandable if one assumes that in the absence of a field the dipoles are in a well-ordered state as may be expected in solids. The field by turning some of the dipoles into different directions can thus only decrease the existing order.

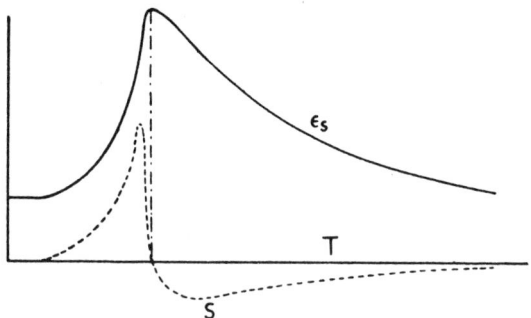

FIG. 3. Schematic temperature-dependence of the dielectric constant ϵ_s and of the entropy change $S \propto \partial\epsilon_s/\partial T$ due to the polarization by a field. If $S > 0$ the field creates disorder, if $S < 0$ it creates order. Near the absolute zero of temperature the substance is already perfectly ordered. Hence $\partial\epsilon_s/\partial T$ cannot be negative near $T = 0$.

B. *Periodic fields*

The corresponding calculations in the general case of a time-dependent field become very involved. It is easy, however, to consider the isothermal case in the presence of a periodic field, and to calculate on an average over one period the amount of electric energy which is transformed into heat.

On an average over a period the energy U of the dielectric cannot change because the temperature is kept constant, and the field E is periodic. Thus in equation 3.5 $dU = 0$, and hence integrating over one period

$$\int dQ = -\frac{1}{4\pi} \int E\,dD = -\frac{1}{4\pi} \int_0^{2\pi/\omega} E \frac{\partial D}{\partial t}\,dt.$$

The heat produced per second and per unit volume, i.e. the

rate of loss L of energy from the electric field, is therefore given by

$$L = \frac{\omega}{8\pi^2} \int_0^{2\pi/\omega} E \frac{\partial D}{\partial t} dt.$$

Inserting here E from equation 2.1 and D from 2.2 and 2.4, one finds

$$L = \frac{\omega E_0^2}{8\pi^2} \int_0^{2\pi/\omega} \cos\omega t \left(\epsilon_1 \frac{\partial \cos\omega t}{\partial t} + \epsilon_2 \frac{\partial \sin\omega t}{\partial t} \right) dt,$$

or after integration,
$$L = \frac{\epsilon_2 E_0^2 \omega}{8\pi}. \tag{3.15}$$

This can be expressed in terms of the phase-shift ϕ (equation 2.5) by

$$L = \frac{\epsilon_1 E_0^2 \omega}{8\pi} \tan\phi. \tag{3.16}$$

For this reason ϕ is usually described as the loss angle.

Another derivation of equation 3.15 is given in the appendix (A 1.ii). The connexion of ϵ_1 and ϵ_2 with the optical constants of the material is also discussed in the appendix (A. 1.iv).

CHAPTER II
STATIC DIELECTRIC CONSTANT

4. Survey

IN the present chapter the intention is to calculate the electric dipole moment induced by an external field in a dielectric from its atomic and molecular structure. The dielectric constant ϵ_s can then be obtained with the help of equation 1.9.

In § 7 a formula of general validity will be derived which connects the static dielectric constant ϵ_s with structural properties of the substance. The calculation of explicit values of ϵ_s and its temperature dependence is usually, however, beset with great formal difficulties and two types of approximations are, therefore, introduced.

Firstly, a simple model is in general chosen to represent a material of much greater complexity. Secondly, mathematical approximations are often introduced which hold only within a certain range of some parameter such as the temperature. Consequently, in order to decide whether an approximate formula is applicable for a given material, one has to judge (a) whether the basic model can be chosen to represent the actual material, and (b) whether the mathematical approximation holds for the given range of the parameter concerned.

A dielectric substance can be considered as consisting of elementary charges e_i, and

$$\sum e_i = 0 \qquad (4.1)$$

if it contains no net charge. The electric dipole moment† of a charge e_i relative to a fixed point is defined by the vector $e_i \mathbf{l}_i$ if $\mathbf{l}_i$ is the radius vector from the fixed point to e_i. The total moment of the whole system is the vector sum of all the individual dipoles, $\sum_i e_i \mathbf{l}_i$. This quantity is independent of the position of the fixed point if the sum of the charges is zero. For using this fact (cf. 4.1) the dipole moment relative to a point at a distance $\mathbf{b}$ from the original one is

$$\sum e_i(\mathbf{l}_i + \mathbf{b}) = \sum e_i \mathbf{l}_i + \mathbf{b} \sum e_i = \sum e_i \mathbf{l}_i.$$

† Also referred to as electric moment or dipole moment or moment.

It will now be assumed that in the lowest energy state (ground level) of the substance its dipole moment vanishes. Then if l_{i0} is the positional vector of the charge e_i in the ground level,

$$\sum_i e_i \mathbf{l}_{i0} = 0. \tag{4.2}$$

Therefore, if $\mathbf{r}_i$ is the displacement of the charge e_i from its equilibrium position in the ground state, $\mathbf{l}_i = \mathbf{l}_{i0} + \mathbf{r}_i$. Then using 4.2,

$$\mathbf{M}(X) = \sum_i e_i \mathbf{l}_i = \sum_i e_i \mathbf{r}_i \tag{4.3}$$

is the electric dipole moment of the substance for a given set of displacements

$$X = (\mathbf{r}_1, \mathbf{r}_2, ..., \mathbf{r}_i, ...) \tag{4.4}$$

which has been expressed in abbreviated form by X. Many such sets X may lead, of course, to the same moment $\mathbf{M}$.

Very often it is useful to collect some of the elementary charges into a group forming an atom, a molecule, a unit cell of a crystal, or some larger unit. Let the jth unit of this type contain the s elementary charges $e_{j1}, e_{j2}, ..., e_{jk}, ..., e_{js}$, and let

$$x_j = (\mathbf{r}_{j1}, \mathbf{r}_{j2}, ..., \mathbf{r}_{jk}, ..., \mathbf{r}_{js}) \tag{4.5}$$

be an abbreviation for the set of all their displacements $\mathbf{r}_{j1}, ..., \mathbf{r}_{js}$.

Then

$$\mathbf{m}(x_j) = \sum_{k=1}^{s} e_{jk} \mathbf{r}_{jk} \tag{4.6}$$

is the electric moment of this jth group of charges, and

$$\mathbf{M}(X) = \sum_j \mathbf{m}(x_j) \tag{4.7}$$

where the sum extends over all the groups. The vector sum of their individual moments $\mathbf{m}(x_j)$ thus forms the total moment $\mathbf{M}(X)$. Our task is to find the average displacements, and hence the average electric moment under the influence of an external electric field.

In order to obtain a preliminary idea about the average contributions of certain displacements to the electric moment we shall consider two cases, each of which has its characteristic type of displacement:

Case (i) the displaced charge is bound elastically to an equilibrium position;

Case (ii) a charge has several equilibrium positions, each of which it occupies with a probability which depends on the strength of an external field.

The interpretation of case (i) is that on displacing the charge e, carried by a particle of mass m a distance $\mathbf{r}$, a restoring force proportional to $-\mathbf{r}$ acts on the particle in a direction opposite to the displacement (hence the $-$ sign). Thus, if a constant external field $\mathbf{f}$ is applied ($t = $ time)

$$\frac{d^2\mathbf{r}}{dt^2} = -\omega_0^2 \mathbf{r} + \frac{e}{m}\mathbf{f}, \qquad (4.8)$$

where $\omega_0/2\pi$ denotes the frequency of oscillation, and $-m\omega_0^2 \mathbf{r}$ is the restoring force. Equation 4.8 can be written

$$\frac{d^2}{dt^2}(\mathbf{r}-\bar{\mathbf{r}}) = -\omega_0^2(\mathbf{r}-\bar{\mathbf{r}}), \qquad (4.9)$$

where
$$\bar{\mathbf{r}} = \frac{e}{m\omega_0^2}\mathbf{f}, \qquad (4.10)$$

i.e. $d\bar{\mathbf{r}}/dt = 0$. The charge e, therefore, carries out harmonic oscillations about the position $\bar{\mathbf{r}}$ which thus represents the time average of its displacement, i.e. if $\mathbf{C}$ and δ are constant

$$\mathbf{r} = \bar{\mathbf{r}} + \mathbf{C}\cos(\omega_0 t + \delta).$$

The average electric moment is, therefore,

$$e\bar{\mathbf{r}} = \frac{e^2}{m\omega_0^2}\mathbf{f}. \qquad (4.11)$$

As an example of case (ii) consider a particle with charge e possessing two equilibrium positions A and B, separated by a distance $\mathbf{b}$. In the absence of an electric field the particle has the same energy in each position. Thus it may be assumed to move in a potential field of the type shown in Fig. 4. If in equilibrium with its surroundings it will oscillate with an energy of order kT about either of the equilibrium positions, say about A. Occasionally, however, through a fluctuation, it will acquire sufficient energy to jump over the potential wall separating it from B. On a time average, therefore, it will stay in A as long as in B, i.e. the probability of finding it in either A or B is $\frac{1}{2}$.

The presence of a field **f** will affect this in two ways. Firstly, as in case (i), the equilibrium positions will be shifted by an amount $\bar{\mathbf{r}}$ which for simplicity will be assumed to be the same in A and B. Secondly, the potential energies V_A, V_B of the particle in the two equilibrium positions will be altered because

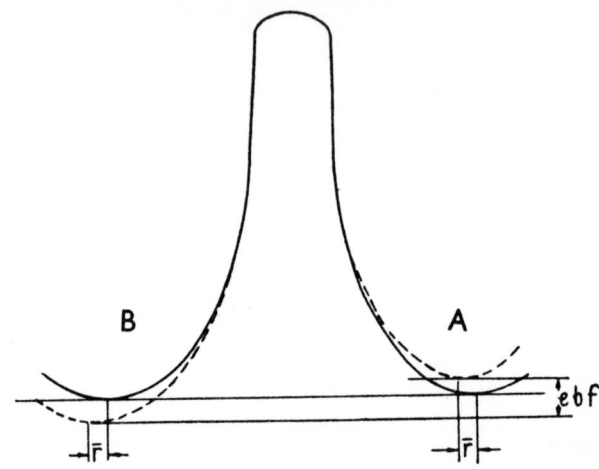

FIG. 4. Potential energy of a charged particle with two equilibrium positions. The dotted curve holds in the presence of an external field **f**.

its interaction energy with the external field differs by $e(\mathbf{bf})$, i.e.

$$V_A - V_B = e(\mathbf{bf}). \qquad (4.12)$$

The particle should, therefore, on the average spend more time near B than near A. Actually, since according to statistical mechanics, the probability of finding a particle with an energy V is proportional to $e^{-V/kT}$,

$$p_A = \frac{e^{-V_A/kT}}{e^{-V_A/kT}+e^{-V_B/kT}}, \qquad p_B = \frac{e^{-V_B/kT}}{e^{-V_A/kT}+e^{-V_B/kT}} \qquad (4.13)$$

are the probabilities for positions A and B respectively. They have been normalized in such a way as to make

$$p_A + p_B = 1 \qquad (4.14)$$

in agreement with the physical condition that the particle must be in one of the two positions. Thus from 4.13 and 4.12

$$p_B - p_A = \frac{e^{e(\mathbf{bf})/kT}-1}{e^{e(\mathbf{bf})/kT}+1} > 0. \qquad (4.15)$$

It follows from the definition of the probabilities p_A and p_B that if the condition of the system over a long time t_1 is considered, the particle will spend a time (use 4.14)

$$p_A t_1 = [\tfrac{1}{2} - \tfrac{1}{2}(p_B - p_A)] t_1$$

in position A, and a time $p_B t_1 = [\tfrac{1}{2} + \tfrac{1}{2}(p_B - p_A)] t_1$ in position B. It has thus been displaced by the distance **b** from A to B during the fraction $\tfrac{1}{2}(p_B - p_A)$ of the time t_1. The average moment induced by the field is thus

$$\tfrac{1}{2} e\mathbf{b}(p_B - p_A). \tag{4.16}$$

Hence if θ is the angle between **b** and **f**, the projection of the induced moment into the field direction is, using 4.16 and 4.15, given by

$$\tfrac{1}{2} eb \cos\theta \frac{e^{ebf \cos\theta/kT} - 1}{e^{ebf \cos\theta/kT} + 1}. \tag{4.17}$$

In most cases it is permissible to assume

$$ebf \ll kT, \tag{4.18}$$

for putting e = electronic charge, $f = 300$ volts/cm. = 1 e.s.u., $b = 10^{-8}$ cm. $\simeq$ distance between neighbouring atoms in a molecule, and $T = 300°$ (= room temperature) one finds

$$\frac{ebf}{kT} \simeq \frac{4.8 \times 10^{-10} \times 10^{-8} \times 1}{1.4 \times 10^{-16} \times 300} \simeq 10^{-4}. \tag{4.19}$$

Developing 4.17 in terms of ebf/kT, the average induced moment in the field direction is found to be

$$(\tfrac{1}{2} eb)^2 \cos^2\theta \, f/kT + e\bar{r}, \tag{4.20}$$

where $e\bar{r}$ is a term similar to those considered in case (i) which has been added to account for the elastic displacement.

Often two charges $+e$ and $-e$ are strongly bound, forming an electric dipole $\boldsymbol{\mu} = e\mathbf{d}$ where **d** is the distance between them. The above case (ii) then leads to the same result as that of a dipole $\boldsymbol{\mu}$ having two equilibrium positions with opposite dipole direction, but with equal energy in the absence of a field. In a field **f** the energy of interaction between field and dipole is given by

$$-(\boldsymbol{\mu}\mathbf{f}), \tag{4.21}$$

so that $2\mu f \cos\theta$ is the energy difference between the two positions

if θ is the angle between μ and f. This is equivalent to equation 4.12 if
$$\mu = \tfrac{1}{2}e\mathbf{b}. \tag{4.22}$$
Actually putting an immobile charge $-e$ half-way between A and B turns case (ii) into the present case. Clearly the induced moment must be the same for both cases because the charge $-e$ is immobile, and its distance from A and B is $\tfrac{1}{2}b$, leading to a dipole moment μ, 4.22. Introducing this into 4.20 yields for the induced moment in the field direction
$$\frac{\mu^2 \cos^2\theta}{kT} f + e\bar{r}. \tag{4.23}$$
In contrast to case (i) the electric moment now depends on temperature. In view of equation 1.9, a substance consisting of a great number of such dipoles will have a temperature-dependent dielectric constant in contrast to a substance in which all charges are bound elastically. According to equation 3.12 this means that in the dipolar case (ii) the entropy of the substance is decreased by the field. This is evident because the field causes the fraction p_B of dipoles with components in the field direction to be larger than the fraction p_A of dipoles with components in the opposite direction, thus leading to a state which is disordered to a smaller degree (i.e. having a lower entropy) than the state of complete disorder in which $p_B = p_A$.

The difference between the action of the field in the two cases (i) and (ii) should be well understood because this is essential for the whole theory of dielectric constant. In case (i) the field exerts a force on an elastically bound charge, thus shifting its equilibrium position. In case (ii) this force of the field on the charge again leads to contributions of type (i) denoted by $e\bar{r}$ in equations 4.20 and 4.23. It would be wrong, however, to assume that the field by this force turns a dipole from one equilibrium position into another. This is effected in a more indirect way because the field slightly alters the probabilities of a jump of a dipole from one equilibrium position to another. This will be described in greater detail in § 9, where the dynamic properties of the present model will be investigated.

It should also be realized that though every charge is displaced elastically (case (i)) the fraction of dipoles turned by a field of reasonable strength (case (ii)) is very small. This fraction is given by $\frac{1}{2}(p_B - p_A)$ which, according to 4.15, 4.18, and 4.19, is of the order 10^{-4} in a field of 300 volts/cm. Even in a field of 100,000 volts/cm. only about 2 per cent. of all dipoles are orientated.

5. Dipolar interaction

In a dielectric two essentially different types of interaction forces should be distinguished. Forces due to chemical bonds, van der Waals attraction, repulsion forces, and others have all such short ranges that usually interaction between nearest neighbours only need be considered. Compared with these forces dipolar interaction forces have a very long range. This can be readily shown as follows.

As indicated previously (cf. § 4) a polarized dielectric can be considered as composed of small regions each having a certain dipole moment, and the total dipole moment of the body is the vector sum of the moments of these regions. Now it is well known from macroscopic theory that the energy per unit volume of a macroscopic specimen depends on its shape (cf. Appendix A 2.iii). This implies that interaction between dipoles must be taken into account even at macroscopic distances and illustrates the great importance of the dipolar interaction forces.

Due to the long range of the dipolar forces an accurate calculation of the interaction of a particular dipole with all the other dipoles of a specimen would be very complicated. However, very good approximation can be made by considering that the dipoles beyond a certain distance, say† a_m, can be replaced by a continuous medium, having the macroscopic dielectric properties of the specimen. Thus the dipole whose interaction with the rest of the specimen we are calculating may be considered as surrounded by a sphere of radius a_m containing a discrete number of particles, beyond which there is a continuous medium. To make this a good approximation the dielectric properties of the whole region within the sphere should be equal

† The suffix m stands for macroscopic.

to those of a macroscopic specimen, i.e. it should contain a sufficient number of molecules to make fluctuations very small.

From this point of view we are thus led to H. A. Lorentz's [*L3*] method for the treatment of dipolar interaction: from a macroscopic specimen select a microscopic spherical region which is sufficiently large to have the same dielectric properties as a macroscopic specimen. The interaction between the dipoles inside the spherical region will then be calculated in an exact way, but for the calculation of their interaction with the rest of the specimen the latter is considered as a continuous medium. To demonstrate this method we shall make use of a very simple model. It consists of a cubic lattice of 'atoms' whose linear dimensions are very small compared with the lattice distance. Each atom consists of a positive charge $+e$ which is rigidly bound to the lattice point, and of a negative charge $-e$ which is bound elastically to it. The force acting on a negative charge when it is displaced while all others are at rest at their respective equilibrium positions will be denoted as the restoring force. When other charges are displaced as well, an additional force will act on the charge due to a change in the interaction between the charges. It will be assumed that electrostatic interaction only exists, and in particular that there are no short-range forces. Furthermore, the temperature will be assumed to be so close to the absolute zero, $T = 0$, that thermal oscillations can be disregarded.

On these assumptions when a macroscopic electric field **E** is set up in the substance all negative charges are displaced by the same amount, say $\bar{\mathbf{r}}$, each forming a dipole

$$\mathbf{m} = (-e)\bar{\mathbf{r}}. \tag{5.1}$$

The field **f** acting on one charge and displacing it against the restoring force is often referred to as the local field, or the inner field; it has to be distinguished from the macroscopic field **E**. Thus if the restoring force is denoted by $-c^2\mathbf{r}$,

$$\bar{\mathbf{r}} = \frac{(-e)}{c^2}\mathbf{f},$$

and hence according to 5.1

$$\mathbf{m} = \frac{e^2}{c^2}\mathbf{f}. \qquad (5.2)$$

To calculate $\mathbf{f}$ according to Lorentz's method, separate the sources of $\mathbf{f}$ into those inside the spherical region and those outside it, making contributions $\mathbf{f}_i$ and $\mathbf{f}_e$ respectively,

$$\mathbf{f} = \mathbf{f}_i + \mathbf{f}_e. \qquad (5.3)$$

Since all induced dipoles are identical, the field $\mathbf{f}$ must be the same at every lattice point. To find $\mathbf{f}_i$ consider therefore the dipole at the centre of the spherical region and calculate its interaction energy I with all the other dipoles of the region. This will depend on $\bar{r}$, and

$$(-e)\mathbf{f}_i = -\operatorname{grad} I(\bar{r}). \qquad (5.4)$$

To calculate this interaction energy I assume the field to be sufficiently weak to make $\bar{r}$ very small compared with the lattice distance a_0. Each dipole $\mathbf{m}$ can then be considered as a point dipole. The electrostatic interaction energy between two parallel dipoles of equal moment is given by

$$\frac{m^2}{l^3}(1 - 3\cos^2\psi), \qquad (5.5)$$

where l is the distance between them and ψ is the angle between $\mathbf{l}$ and $\mathbf{m}$. For simplicity let $\mathbf{m}$ be parallel to a crystal axis, say in the z-direction. Then the three components of $\mathbf{l}$ are

$$X = na_0, \qquad Y = pa_0, \qquad Z = qa_0$$

where n, p, q are positive or negative integers. Since $\cos\psi = Z/l$, the total energy of interaction follows from 5.5 by summation over all lattice points, i.e.

$$I = m^2 \sum \frac{l^2 - 3Z^2}{l^5} = \frac{m^2}{a_0^3} \sum_{n,p,q} \frac{n^2 + p^2 - 2q^2}{(n^2 + p^2 + q^2)^{\frac{5}{2}}}.$$

Now to each set of three values n, p, q, say $n = n_0$, $p = p_0$, $q = q_0$, two others can be coordinated by cyclic permutation, namely $n = p_0$, $p = q_0$, $q = n_0$ and $n = q_0$, $p = n_0$, $q = p_0$. These three terms just cancel in the above sum, and hence $I = 0$. It thus follows, using 5.4, that

$$f_i = 0. \qquad (5.6)$$

The external contribution f_e has to be calculated macroscopically. It represents the electric field inside the spherical region due to all sources except the polarization inside this region. To understand this clearly it should be remembered that the macroscopic electric field **E** is partly due to true charges outside the specimen (or at its surface) and partly due to the polarization **P** of the dielectric which acts in the opposite direction. To obtain $\mathbf{f}_e$ the contribution to **E** due to the spherical region should be omitted. Therefore if $\mathbf{E}_s$ is this contribution,

$$\mathbf{f}_e = \mathbf{E} - \mathbf{E}_s. \tag{5.7}$$

By definition, $\mathbf{E}_s$, the self-field, is the field inside a spherical specimen with a permanent polarization **P**. Its direction is opposite to **P**. A simple electrostatic calculation (Appendix A 2.21) shows that

$$\mathbf{E}_s = -\frac{4\pi}{3}\mathbf{P}. \tag{5.8}$$

Therefore, using 5.3, 5.6, 5.7, 5.8, and 1.9,

$$\mathbf{f} = \mathbf{f}_e = \mathbf{E} + \frac{4\pi}{3}\mathbf{P} = \frac{\epsilon_s + 2}{3}\mathbf{E}. \tag{5.9}$$

It should be noted that this expression is independent of the size of the spherical region, which means that the contribution of any homogeneously polarized spherical shell to the field inside it vanishes. In other words, the interaction between such a shell and a dipole inside it vanishes. This is in agreement with the above result 5.6 obtained by considering dipole-dipole interaction in detail. The present model, therefore, has been chosen in such a way as to make macroscopic and microscopic treatment identical. Such a result, of course, could only be obtained by assuming short-range forces to be entirely absent. The existence of such a force would modify $\mathbf{f}_i$, but it would not alter our result for $\mathbf{f}_e$.

The fact that the interaction energy of the dipoles of a spherical region vanishes must mean that if a spherical specimen is brought into a homogeneous field $\mathbf{E}_0$ in vacuum this field represents the local field acting on each dipole. Actually from electrostatics it follows that the field inside a spherical specimen of dielectric constant ϵ_s is $\mathbf{E} = 3\mathbf{E}_0/(\epsilon_s+2)$ (cf. Appendix A 2.16), and hence with equation 5.9, $\mathbf{f} = \mathbf{E}_0$.

The above result 5.9 can be derived by an alternative method due to Onsager [*O1*] which is capable of a generalization and which will be of use at a later stage of the development of a general theory.

The reader should realize, however, that in Onsager's paper the spherical region contains one single dipolar molecule only. This is, of course, open to the objection that the spherical region is certainly not large enough to have the same properties as a macroscopic specimen. In our use of his analysis, however, the spherical region is sufficiently large to make it certain that it is correct to treat the outside macroscopically. Now consider, in the absence of a macroscopic field, a spherical region which is polarized homogeneously, thus having a dipole moment $\mathbf{M} = \mathbf{P}V$ where V is its volume. If considered separately without the rest of the specimen the electric field inside the spherical region is the self-field $\mathbf{E}_s$, equation 5.8. If, on the other hand, the spherical region is considered inside the specimen there will be a certain interaction between the polarized sphere and the surroundings, giving rise to an altered field inside the spherical region. Its difference from $\mathbf{E}_s$ is denoted as the reaction-field $\mathbf{R}$ and is given by (Appendix A 2.18; a_m = radius of the spherical region)

$$\mathbf{R} = \frac{2(\epsilon_s-1)}{2\epsilon_s+1}\frac{\mathbf{M}}{a_m^3} = \frac{2}{3}\frac{\epsilon_s-1}{2\epsilon_s+1}\frac{4\pi\mathbf{M}}{V} = \frac{2}{3}\frac{\epsilon_s-1}{2\epsilon_s+1}4\pi\mathbf{P}. \tag{5.10}$$

Thus $\mathbf{R}$ is the field produced inside the spherical region by the surroundings if polarized by the former. If a macroscopic field $\mathbf{E}$ is now produced in the specimen without altering the moment $\mathbf{M}$ of the sphere, the field inside it will be increased by an additional field, the cavity-field $\mathbf{G}$ (cf. Appendix A 2.15),

$$\mathbf{G} = \frac{3\epsilon_s}{2\epsilon_s+1}\mathbf{E}. \tag{5.11}$$

Thus the total field inside the sphere, due to outside sources, is

$$\mathbf{f}_e = \mathbf{G}+\mathbf{R} = \frac{3\epsilon_s}{2\epsilon_s+1}\mathbf{E} + \frac{2}{3}\frac{\epsilon_s-1}{2\epsilon_s+1}4\pi\mathbf{P}. \tag{5.12}$$

This expression holds whatever the value of **P**. In our particular case if $4\pi\mathbf{P} = (\epsilon_s - 1)\mathbf{E}$ (cf. § 1) is introduced, the above expression becomes identical with 5.9.

Finally to calculate ϵ_s, insert **f** from 5.9 into 5.2. The polarization **P** is then given by

$$\mathbf{P} = N_0\mathbf{m} = \frac{\epsilon_s + 2}{3}\frac{e^2}{c^2}N_0\mathbf{E},$$

where N_0 is the number of particles per unit volume. Hence using 1.9

$$\frac{\epsilon_s - 1}{\epsilon_s + 2} = \frac{4\pi}{3}N_0\frac{e^2}{c^2}, \tag{5.13}$$

which is usually known as Clausius–Mossotti formula (cf. Clausius, *C1*; Mossotti, *M5*). The above derivation of this formula is exact except for the assumption that $\bar{r}$ is a small quantity. It will be noted, however, that the change in ϵ_s as $\bar{r}$ increases is essentially a manifestation of the field dependence of the dielectric constant (since the change in magnitude of $\bar{r}$ depends only on field strength) and this does not concern us here. That is, formula 5.13 is correct for the limiting case in which the dielectric constant is independent of the field. The reader should realize, however, that this derivation holds for the above model only; there are few substances which it could claim to represent, though for simple non-polar substances it will often be a useful approximation.

6. Dipolar molecules in gases and dilute solutions

Molecules can be divided into two classes, polar and non-polar, according to whether or not they possess an electric dipole moment in their lowest energy-level (ground state). In general it is easy to recognize the class to which a molecule belongs because a non-polar molecule must have a point of symmetry defined in such a way that the distribution of charges along (or near) any straight line passing through it must be symmetrical with respect to this point. Thus a diatomic molecule is polar unless its two atoms are equal (e.g. the polar molecules HCl, CO, and the non-polar molecules H_2, O_2). Triatomic molecules of the type AB_2, where A and B represent different atoms, are polar unless their nuclei lie on a straight line with A half-way between the two B atoms. Examples are the triangular polar H_2O and the straight non-polar CO_2 molecules. C_6H_6 is a more complicated non-polar molecule forming a plane hexagon with the centre as point of symmetry. On replacing one H atom by another type of atom, say by Cl, the resultant molecule C_6H_5Cl becomes polar.

The magnitude of molecular dipoles is usually of the order of one electronic charge ($\sim 4\cdot 8 \times 10^{-10}$ e.s.u.) displaced by $\frac{1}{4} \times 10^{-8}$ cm., i.e. about 10^{-18} c.g.s. units. Often dipole moments are expressed in Debye units, one such unit being 10^{-18} c.g.s. units. Debye was the first to recognize the importance of the investigation of dipole moments for a study of the constitution of molecules (cf. his book, reference $D2$). A detailed discussion of molecular structure is, however, not intended in the present book. Our purpose in this section is to determine the properties of dielectric substances consisting of molecules having permanent dipole moments and our model for such a molecule (in vacuum and free from perturbing influences) will, therefore, have a dipole moment μ_v.

In addition to its translational motion, a free molecule can carry out oscillations and rotations. However, unless stated otherwise, it will be assumed that these do not alter the average value of the dipole moment. In a complicated molecule intra-molecular rotation of dipolar groups may lead to considerable changes in the dipole moment of the molecule as a whole (cf. the book by Le Fèvre, $L2$). Such molecules will not be considered at present.

A constant electric field $\mathbf{f}$ will influence the molecule in two ways. Firstly, it will perturb the free rotation of the dipole, and secondly, it will induce a further dipole moment, say $\alpha\mathbf{f}$, by elastic displacement of the atomic electrons relative to their respective nuclei, and to a smaller extent by elastic displacement of the nuclei relative to each other. The total moment of the molecule is thus

$$\mathbf{m} = \mathbf{\mu}_v + \alpha\mathbf{f}. \tag{6.1}$$

The quantity α has the dimensions of a volume and is called the polarizability of the molecule. In anisotropic molecules (that is, molecules having different polarizabilities along different axes) the induced moment need not always have the same direction as the field **f**; the polarizability α is in this case not a scalar but a tensor quantity. Its average value, $\bar{\alpha}$, obtained by allowing the molecule to have all possible directions relative to the field, is, however, a scalar. Assuming that the polarizability of a molecule is entirely due to electronic contributions, α is determined by the optical refractive index n. For, according to Maxwell's relation, n^2 is the dielectric constant at optical frequencies at which there are no dipolar contributions (because the time required by the dipoles to attain equilibrium with the field is much longer than the period of the field). Assuming the molecules to be isotropic, it will be shown below that the Clausius–Mossotti formula holds approximately in this case, i.e.

$$\alpha = a^3 \frac{n^2-1}{n^2+2}, \tag{6.2}$$

where a is the radius of a sphere which on an average contains one molecule.

Gases

We shall now proceed to calculate the dielectric constant ϵ_s of a gas of dipolar molecules. In order to simplify calculations, it will be assumed that the density of the gas is so low that the dipolar interaction energy is small compared with the thermal energy ($\simeq kT$ per molecule) and can therefore be neglected.

According to § 5 the dipolar interaction energy is of the order $\mu_v^2/l^3 \simeq \mu_v^2 N_0$ if N_0 is the number of molecules per unit volume. It will thus be assumed that

$$\mu_v^2 N_0 \ll kT. \tag{6.3}$$

This simplifies the calculation of ϵ_s considerably because it implies that the local field **f**, acting on a dipole, is entirely due to external sources, i.e. $\mathbf{f} = \mathbf{D} = \epsilon_s \mathbf{E}$ (cf. § 1). Also if N_0 is small $\epsilon_s - 1$ must be small as well, so that the further assumption

$$\epsilon_s - 1 \ll 1 \tag{6.4}$$

leads to $\qquad\qquad\qquad \mathbf{f} = \mathbf{E}. \tag{6.5}$

It will be found below that 6.4 follows from 6.3.

In order to calculate the electric moment **M** of the gas we make use of the fact (§ 4) that **M** is equal to the vector sum of the moments **m** of all N molecules. Since, on the other hand, the time average $\bar{\mathbf{m}}$ of **m** is the same for all molecules,

$$\mathbf{M} = N\bar{\mathbf{m}}. \tag{6.6}$$

Thus from 1.9
$$\epsilon_s - 1 = 4\pi N_0 \bar{m}/E, \tag{6.7}$$

where according to 6.1 and 6.5

$$\bar{\mathbf{m}} = \bar{\boldsymbol{\mu}}_v + \bar{\alpha}\mathbf{E}. \tag{6.8}$$

In this expression the first term represents the average value of the intrinsic moment, and the second one is the average induced moment per molecule. The calculation of $\bar{\mathbf{m}}$ is similar to that carried out in case (ii) of § 4. In the present case, however, instead of having only two possible directions as described in § 4, a dipole has a whole continuum of possible directions. As in § 4, the behaviour of the dipole may be considered on a statistical basis, without inquiring into its dynamic properties under the influence of the field E. Thus since $-E\mu_v \cos\theta$ is the energy of the dipole in the field, the probability of finding μ_v in a direction forming an angle between θ and $\theta + d\theta$ with **E** is given, according to statistical mechanics, by ($2\pi \sin\theta\, d\theta$ is the element of the solid angle between θ and $\theta + d\theta$)

$$e^{E\mu_v \cos\theta/kT} \sin\theta\, d\theta \bigg/ \int_0^\pi e^{E\mu_v \cos\theta/kT} \sin\theta\, d\theta, \tag{6.9}$$

provided that thermal equilibrium has been attained. At present in considering static properties this will always be assumed to be the case.

In evaluating 6.9 we shall assume, as in the case of equation 4.18, that the field is sufficiently weak to make

$$\frac{\mu_v E}{kT} \ll 1. \tag{6.10}$$

Then developing 6.9 in terms of $\mu_v E/kT$, and keeping terms up

to the first order only, one obtains for the average value of†
$\cos\theta$,

$$\overline{\cos\theta} = \int_0^\pi \cos\theta e^{E\mu_v \cos\theta/kT}\sin\theta\,d\theta \Big/ \int_0^\pi e^{E\mu_v \cos\theta/kT}\sin\theta\,d\theta = \frac{\mu_v E}{3kT}.$$
(6.11)

Similarly the components of $\mathbf{\mu}_v$ perpendicular to $\mathbf{E}$ are found to vanish, so that the average moment $\bar{\mathbf{\mu}}_v$ of a molecule is a vector with the same direction as $\mathbf{E}$, amounting to $\bar{\mu}_v = \mu_v \overline{\cos\theta}$, i.e.

$$\bar{\mathbf{\mu}}_v = \frac{\mu_v^2}{3kT}\mathbf{E}. \qquad (6.12)$$

On using 6.7 and 6.8, therefore,

$$\epsilon_s - 1 = \frac{4\pi\mu_v^2 N_0}{3kT} + 4\pi\bar{\alpha}N_0, \quad \text{if } \epsilon_s - 1 \ll 1 \qquad (6.13)$$

or introducing $\qquad \epsilon_\infty = 1 + 4\pi\bar{\alpha}N_0 \qquad (6.14)$

as dielectric constant at frequencies so high that the dipoles have no time to attain equilibrium,

$$\epsilon_s - \epsilon_\infty = \frac{4\pi\mu_v^2 N_0}{3kT}, \quad \text{if } \epsilon_s - 1 \ll 1. \qquad (6.15)$$

In agreement with the conclusions drawn in § 4 the dipolar contribution to ϵ_s is temperature-dependent in contrast to the non-polar contribution ϵ_∞. Measurement of the temperature-dependence of ϵ_s thus makes possible the separation of the dipolar contribution, $\epsilon_s - \epsilon_\infty$, and hence the determination of μ_v (examples in § 16).

Dilute solutions

The above derivation suggests that a formula similar to 6.15 should hold for dilute solutions of dipolar molecules in a non-polar liquid having the temperature-independent dielectric constant ϵ_0. For at sufficiently low densities the interaction between dipoles can again be neglected. Also, replacing 6.4 by

$$\epsilon_s - \epsilon_0 \ll 1, \qquad (6.16)$$

† For with $x = \cos\theta$ and $\gamma = \mu_v E/kT \ll 1$,

$$\overline{\cos\theta} = \int_{-1}^1 x e^{\gamma x}\,dx \Big/ \int_{-1}^1 e^{\gamma x}\,dx \simeq \int_{-1}^1 (x + \gamma x^2)\,dx \Big/ \int_{-1}^1 dx = \frac{0 + 2\gamma/3}{2} = \gamma/3.$$

equation 6.5 holds as before, leading to 6.11 because in a liquid, in the absence of an external field, all directions of the dipole are equally probable. Furthermore, if ϵ_∞ is again the non-polar contribution to ϵ_s, then $(\epsilon_\infty-1)E/4\pi$ represents the non-dipolar contribution to the electric moment of the solution per unit volume. Since $(\epsilon_s-1)E/4\pi$ is the total moment, $(\epsilon_s-\epsilon_\infty)E/4\pi$, in view of the additivity of moments, is, as before, the dipolar contribution. Assuming that the dipole is rigid, i.e. that $\alpha = 0$, equation 6.15 would follow with $\epsilon_\infty \simeq \epsilon_0$. Actually a molecular dipole is not rigid, but can be polarized. In a solution this leads to a change in the effective dipole moment of the molecule, as was recognized by various authors (cf. Weigle, *W1*; Frank, *F1*; Higasi, *H2*; Frank and Sutton, *F4*; these authors also consider other effects which may lead to an alteration in the effective dipole moment when the molecule is non-spherical or has a large quadripole moment). The dipole polarizes its surroundings which in turn produce a reaction field at the position of the dipole. This field polarizes the molecule and thus alters its dipole moment. Thus if we define the resultant moment which the molecule has in the solution as the 'internal moment' μ_i, the reaction field will be proportional to μ_i, say $g\mu_i$. In the absence of an external field, therefore, from 6.1 (using $\mathbf{f} = g\mu_i$ and $\mathbf{m} = \mu_i$)

$$\mu_i = \mu_v + \alpha g \mu_i, \tag{6.17}$$

or
$$\mu_i = \frac{\mu_v}{1-\alpha g}. \tag{6.18}$$

A quantitative calculation of the reaction field, and hence of μ_i, is very difficult, however, except in the case of spherical molecules.

It follows from the above that, when a field $\mathbf{E}$, produced by external sources, exists in the solution the resultant field acting upon a molecule in the solution will be different from $\mathbf{E}$. Also the moment μ_i of the molecule in solution will be different from that it has in vacuum (namely μ_v). In order to take account of these facts in calculating the dielectric constant of the solution we may consider the dipole in one of two ways. In the first of these we calculate the field acting upon the 'internal moment'

μ_i. This will, of course, depend on the shape chosen for our model of the molecule. In the second method we calculate the interaction between the field of the dipolar molecule in the solution around it and the applied field **E**. The same dipolar field might alternatively be produced by a rigid dipole having in vacuum a moment μ_e which will be denoted as the 'external moment' of the molecule.† A solution of rigid dipoles μ_e should, therefore, have the same dielectric constant as the solution which we actually consider.

The former method will be used below in deriving the Onsager formula. At present the latter method will be chosen because it enables us to make use of our previous derivation for gases to determine the dielectric constant ϵ_s of the solution, i.e. to employ equation 6.15 after replacing μ_v by μ_e. Thus with the use of 6.18

$$\epsilon_s - \epsilon_\infty = \frac{4\pi\mu_e^2 N_0}{3kT} = \frac{4\pi\mu_v^2 N_0}{3kT}\left(\frac{\mu_e/\mu_i}{1-\alpha g}\right)^2,$$

if $\quad\quad \epsilon_s - \epsilon_\infty \ll 1 \quad$ and $\quad \epsilon_\infty - \epsilon_0 \ll 1. \quad\quad$ (6.19)

Here the ratio μ_e/μ_i of external to internal moment depends on the shape of the molecule. For spherical molecules, according to the definitions of μ_e and μ_i, the internal moment μ_i is identical with the moment of a sphere inside the dielectric containing μ_e at its centre. An ordinary calculation in electrostatics shows that in this case (cf. Appendix A 2.31)

$$\mu_e = \frac{3\epsilon_s}{2\epsilon_s+1}\mu_i, \tag{6.20}$$

and if a is the molecular radius (cf. Appendix A 2.19)

$$g = \frac{2(\epsilon_s-1)}{2\epsilon_s+1}\frac{1}{a^3}. \tag{6.21}$$

Furthermore, if the polarizability is isotropic and is mainly due to displacement of electrons, equation 6.2 holds, n being the refractive index of a pure liquid of the dipolar molecule. Then with 6.18, using 6.21 and 6.2, the internal moment of a spherical molecule is

$$\mu_i = \left(\frac{2\epsilon_s+1}{2\epsilon_s+n^2}\frac{n^2+2}{3}\right)\mu_v, \tag{6.22}$$

† Compare Appendix A 2.ii for a discussion of dipole moments.

and with 6.20 its external moment is

$$\mu_e = \frac{\epsilon_s(n^2+2)}{2\epsilon_s+n^2}\mu_v. \quad (6.23)$$

The dielectric constant in this case is (cf. 6.19) given by (replacing ϵ_s by ϵ_0 on the right-hand side according to 6.16)

$$\epsilon_s - \epsilon_\infty = \frac{4\pi\mu_v^2 N_0}{3kT}\left(\frac{\epsilon_0(n^2+2)}{2\epsilon_0+n^2}\right)^2 \quad (6.24)$$

$$= \frac{4\pi\mu_v^2 N_0}{3kT}\left(\frac{\epsilon_0+2}{3}\right)^2\left\{1-\frac{2(\epsilon_0-1)(\epsilon_0-n^2)}{(2\epsilon_0+n^2)(\epsilon_0+2)}\right\}^2$$

for spherical molecules and $\epsilon_s - \epsilon_\infty \ll 1$.

In some cases we might expect the approximation of spherical molecules to give a correct order of magnitude for the deviation from unity of the ratio of the dipole moments in vacuum and in solution, but we should not expect to obtain a more accurate result than that.

Onsager formula

In the case of spherical molecules Onsager [*O1*] has shown that it is possible to go one step farther in the approximate calculation of the dielectric constant. The interaction between molecules will no longer be entirely neglected, but one component of it, namely the long-range dipolar interaction, will be taken into account. The following assumptions will, therefore, be made:

(a) A molecule occupies a sphere of radius a and its polarizability is isotropic;

(b) The short range interaction energy is negligible (i.e. $\ll kT$ per molecule).

Assumption (b) means that the surroundings of a molecule will be treated as a macroscopic continuum with dielectric constant ϵ_s because long-range forces only will be considered. Estimates on the range of validity of this method can be made (cf. reference *F12*), but will be postponed until the more general Kirkwood formula is derived (§ 8).

On the assumptions made above, the contribution of a single molecule to the dielectric constant ϵ_s can be calculated in the same way for a pure liquid or for a mixture. In both cases the local field **f** acting on the molecule is the field inside a cavity of radius a within a continuous medium of dielectric constant ϵ_s, and is composed of (i) the cavity field **G** due to external sources, and (ii) of the reaction field **R** due to the moment **m** of the molecule itself. Both **G** and **R** were discussed in § 5, although there the cavity of radius d contained a great number of molecules. To use the same values for **G** and **R** in the present case is only possible because of the assumptions (a) and (b). Thus, according to 5.10, 5.11, and 5.12 (**M**/a_m^3 of § 5 corresponds now to **m**/a^3) and using 6.21

$$\mathbf{f} = \mathbf{G} + \mathbf{R} = \frac{3\epsilon_s}{2\epsilon_s+1}\mathbf{E} + g\mathbf{m}. \tag{6.25}$$

Inserting this value into 6.1 the moment **m** becomes

$$\mathbf{m} = \boldsymbol{\mu}_v + \frac{3\epsilon_s}{2\epsilon_s+1}\alpha\mathbf{E} + \alpha g\mathbf{m},$$

or solving with respect to **m**,

$$\mathbf{m} = \frac{\boldsymbol{\mu}_v}{1-\alpha g} + \frac{3\epsilon_s}{2\epsilon_s+1}\frac{\alpha\mathbf{E}}{1-\alpha g}. \tag{6.26}$$

With this value for **m** the internal field **f** (6.25) becomes

$$\mathbf{f} = \frac{3\epsilon_s}{2\epsilon_s+1}\frac{\mathbf{E}}{1-\alpha g} + \frac{g}{1-\alpha g}\boldsymbol{\mu}_v. \tag{6.27}$$

To calculate the average polarization $\overline{\mathbf{m}}$ we shall require $\overline{\boldsymbol{\mu}}_v$. Considering that

$$-\boldsymbol{\mu}_v\mathbf{f} = -\frac{3\epsilon_s}{2\epsilon_s+1}\frac{E\mu_v\cos\theta}{1-\alpha g} - \frac{g}{1-\alpha g}\mu_v^2 \tag{6.28}$$

is the energy of a dipole $\boldsymbol{\mu}_v$ in the field **f**, the probability of finding it in a direction forming an angle θ with **E** is (similar to 6.9) given by

$$e^{\boldsymbol{\mu}_v\mathbf{f}/kT}\sin\theta\,d\theta \Big/ \int_0^\pi e^{\boldsymbol{\mu}_v\mathbf{f}/kT}\sin\theta\,d\theta, \tag{6.29}$$

where $\boldsymbol{\mu}_v\mathbf{f}$ should be introduced from 6.28. It will be seen that the second term in 6.28 is independent of θ and therefore in 6.29

only the first term remains. Following the same procedure as in the derivation of 6.11 one finds

$$\overline{\cos\theta} = \frac{3\epsilon_s}{2\epsilon_s+1} \frac{\mu_v E}{3kT(1-\alpha g)},$$

and hence
$$\bar{\mu}_v = \frac{3\epsilon_s}{2\epsilon_s+1} \frac{\mu_v^2}{3kT(1-\alpha g)} \mathbf{E}. \qquad (6.30)$$

Thus from 6.26, using 6.30 (and $\bar{\alpha} = \alpha$ for isotropic polarization), the average moment $\overline{\mathbf{m}}$ of the molecule is given by

$$\overline{\mathbf{m}} = \frac{3\epsilon_s}{2\epsilon_s+1}\left(\frac{\mu_v^2}{3kT(1-\alpha g)^2} + \frac{\alpha}{1-\alpha g}\right)\mathbf{E}. \qquad (6.31)$$

Let us consider first the application of 6.31 to the simple case of a pure liquid of non-polar molecules for which $\mu_v = 0$. Since by definition a molecule occupies an average volume of

$$\frac{4\pi}{3}a^3 = \frac{1}{N_0}, \qquad (6.32)$$

equations 6.7 and 6.31 yield

$$\epsilon_s - 1 = \frac{3\epsilon_s}{2\epsilon_s+1} \frac{\alpha}{1-\alpha g} \frac{3}{a^3}. \qquad (6.33)$$

Introducing g from 6.21 and solving with respect to α leads to the Clausius–Mossotti equation

$$\frac{\epsilon_s-1}{\epsilon_s+2} = \frac{\alpha}{a^3}. \qquad (6.34)$$

This equation can be considered to prove 6.2, which, together with 6.21, transforms 6.31 into

$$\overline{\mathbf{m}} = \frac{3\epsilon_s \mathbf{E}}{2\epsilon_s+n^2}\left\{\frac{2\epsilon_s+1}{2\epsilon_s+n^2}\left(\frac{n^2+2}{3}\right)^2 \frac{\mu_v^2}{3kT} + \frac{n^2-1}{3}a^3\right\}. \qquad (6.35)$$

The Onsager formula for a pure dipolar liquid is then obtained by inserting 6.35 into 6.7, making use of 6.32:

$$\epsilon_s - n^2 = \frac{3\epsilon_s}{2\epsilon_s+n^2} \frac{4\pi\mu_v^2 N_0}{3kT}\left(\frac{n^2+2}{3}\right)^2. \qquad (6.36)$$

For very small densities this expression becomes identical with 6.15, as should be expected (because for $\epsilon_s-1 \ll 1$ and $n^2-1 \ll 1$, $(n^2+2)/3 \simeq 1$ and $3\epsilon_s/(2\epsilon_s+1) \simeq 1$).

Consider next a mixture containing $N_1, N_2,..., N_s,..., N_z$ molecules per c.c. of z different compounds. Then $\sum_{s=1}^{z} N_s \overline{\mathbf{m}}_s$ is the total moment per c.c. where $\overline{\mathbf{m}}_s$ is the average moment of a molecule of the type s obtained from 6.35 by inserting for μ_v, n, and a the respective values $\mu_{vs}, n_s,$ and a_s which these quantities have for the sth kind of molecule. From 1.9 the dielectric constant of the mixture thus follows:

$$\epsilon_s - 1 = 4\pi \sum_{s=1}^{z} N_s \overline{m}_s / E. \tag{6.37}$$

As a particular case take a mixture containing per c.c. N_1 polar (μ_{v1}, n_1, a_1) and N_2 non-polar ($\mu_{v2} = 0, n_2, a_2$) molecules. Inserting $\overline{m}_1$ and $\overline{m}_2$ from 6.35, equation 6.37 leads to

$$\epsilon_s - 1 = \frac{3\epsilon_s}{2\epsilon_s+1} \left\{ \frac{4\pi \mu_{v1}^2 N_1}{3kT} \left(\frac{2\epsilon_s+1}{2\epsilon_s+n_1^2} \frac{n_1^2+2}{3} \right)^2 + \right.$$
$$\left. + \frac{2\epsilon_s+1}{2\epsilon_s+n_1^2}(n_1^2-1)\frac{4\pi}{3}a_1^3 N_1 + \frac{2\epsilon_s+1}{2\epsilon_s+n_2^2}(n_2^2-1)\frac{4\pi}{3}a_2^3 N_2 \right\}. \tag{6.38}$$

For a dilute solution of polar molecules, i.e. when $N_1 \ll N_2$, this expression should become identical with 6.24 if $n = n_1$, $n_2^2 = \epsilon_0$, and $N_1 = N_0$. Actually since $4\pi a_2^3 N_2/3$ represents the volume occupied by the N_2 non-polar molecules in 1 c.c., $N_1 \ll N_2$ means that $4\pi a_2^3 N_2/3 \simeq 1$. Since the second term in 6.38 can be neglected, we find with the above substitutions for $n_1, n_2,$ and N_1,

$$\epsilon_s - 1 = \frac{3\epsilon_s}{2\epsilon_s+\epsilon_0}(\epsilon_0-1) + \frac{3\epsilon_s}{2\epsilon_s+1} \frac{4\pi \mu_{v1}^2 N_0}{3kT}\left(\frac{2\epsilon_s+1}{2\epsilon_s+n^2}\frac{n^2+2}{3}\right)^2,$$

which, after a simple transformation, becomes identical with 6.24 (since $\epsilon_s \simeq \epsilon_0 \simeq \epsilon_\infty$).

7. General theorems [F10]

The formulae derived in the previous section for the dielectric constant ϵ_s have the great advantage of representing ϵ_s in a very simple way with the help of a few parameters. It should not be forgotten, however, that they only hold subject to certain conditions which in many practically important cases are not fulfilled. In the present section, therefore, expressions for the static dielectric constant ϵ_s will be derived which hold in a very general way for any dielectric substance which is not permanently polarized. Clearly the derivation of such expressions requires some mathematical abstractions, but in view of their importance it was thought desirable to give a detailed account of their deduction.

As in § 5, select from an infinite homogeneous specimen a macroscopic spherical region of volume V which is large compared with a region just big enough to have the same dielectric properties as a macroscopic specimen. The surface of the spherical region need not be an exact geometrical sphere; deviations of a molecular magnitude are permitted and the exact surface will be laid in such a way that no molecules are cut by the surface. This has no influence on the field at distances which are large compared with atomic dimensions. We shall calculate the projection M_E of the average electric moment of the sphere in the direction of the macroscopic field $\mathbf{E}$. For this purpose all particles inside the sphere will be treated according to the rules of classical statistical mechanics. The outside, however, will be considered as a continuous dielectric described by the macroscopic dielectric constant ϵ_s. It will be assumed throughout that the macroscopic field E is sufficiently weak to prevent saturation, so that ϵ_s is independent of E.

The spherical region consists of a number of elementary charges e_i, each of which can be described in terms of its displacement from the position it would have in the lowest energy-level (ground state) of the whole system. This displacement is a vector quantity and is denoted by $\mathbf{r}_i$. A set of all the displacement vectors will be collectively denoted by X, according to equation 4.4. Except at the absolute zero of temperature, such a system of particles does not stay in the same configuration, even if macroscopically it is in equilibrium. Owing to thermal fluctuations there is a probability

$$e^{-U(X,E)/kT}\,dX \Big/ \int e^{-U(X,E)/kT}\,dX \qquad (7.1)$$

of finding it with any set of displacements in a space element between

$X = (\mathbf{r}_1, \mathbf{r}_2, ..., \mathbf{r}_i, ...)$ and $X + dX = (\mathbf{r}_1 + d\mathbf{r}_1, ..., \mathbf{r}_i + d\mathbf{r}_i, ...)$.

Here, $U(X, E)$ is the potential energy of the system in the configuration X, in the presence of a field E, and

$$dX = d\mathbf{r}_1 d\mathbf{r}_2..., d\mathbf{r}_i... \qquad (7.2)$$

is the product of the volume elements $d\mathbf{r}_i$ of all the displacements $\mathbf{r}_i$, each being again a product of its components $dr_{ix} dr_{iy} dr_{iz}$. The integration must be carried out over all possible values of all the displacements.

Each set of displacements X leads to a dipole moment $\mathbf{M}(X)$ as shown in equation 4.3. Therefore, if θ is the angle between $\mathbf{M}(X)$ and $\mathbf{E}$, $M(X)\cos\theta$ represents the projection of $\mathbf{M}(X)$ in the direction of the macroscopic electric field $\mathbf{E}$, and M_E is the average value of $M(X)\cos\theta$. Thus using 7.1,

$$M_E = \int M(X)\cos\theta \, e^{-U(X,E)/kT} dX \Big/ \int e^{-U(X,E)/kT} dX. \qquad (7.3)$$

The value $U(X, 0)$ of the energy in the absence of a macroscopic field will be denoted by $U(X)$, and the zero of potential energy will be chosen so that $U(X)$ vanishes in the ground state, i.e. when all displacements vanish. The energy $U(X)$ can be considered as composed of (i) the energy of interaction $U_i(X)$ between the particles of the spherical region, and of (ii) their energy of interaction $U_e(X)$ with the external region,

$$U(X) = U_i(X) + U_e(X). \qquad (7.4)$$

$U_e(X)$ will depend not only on X, but also on the dielectric constant ϵ_s because the external region is to be treated on a macroscopic basis.

The fact that $U_e(X)$ contains the parameter ϵ_s which may depend on temperature requires some consideration. Our present investigation is a special case of a wider group of problems—the statistical mechanics of systems containing temperature-dependent parameters—which has been investigated by Gross and Halpern [G1]. Applied to our case their investigations show that ϵ_s has to be treated as constant parameter whenever differentiations with respect to temperature are required; $U_e(X)$ is then the energy required to establish at constant temperature the equilibrium polarization in the outside region. From the point of view of the whole system (spherical region

+outside region) $U_e(X)$ is thus the corresponding free energy composed of the energy of interaction of the charges inside the sphere with the polarization induced by them in the outside region, and of the free energy of this region (so far as it depends on X), as shown in the subsequent example (cf. 7.17 and 7.18). This is reasonable because $U(X)$ is the energy required to establish the displacement X at constant temperature.

Now for a given configuration X, assume the macroscopic field **E** to be applied (which, of course, alters the probability of this configuration). Inside the spherical region this leads to an additional homogeneous field **G** (the cavity field) given by equation 5.11. It follows from electrostatics that its interaction with the charges of the spherical region is given by

$$-\mathbf{M}(X)\mathbf{G} = -\frac{3\epsilon_s}{2\epsilon_s+1}M(X)E\cos\theta. \qquad (7.5)$$

Thus in the presence of a field E,

$$U(X,E) = U(X) - \frac{3\epsilon_s}{2\epsilon_s+1}M(X)E\cos\theta. \qquad (7.6)$$

This expression for $U(X, E)$ must be inserted into equation 7.3. It should be remembered now that E was to be considered as sufficiently weak to prevent saturation. Thus if the right-hand side of 7.3 is developed into a power series in E, only the first term need be considered, i.e.

$$e^{-U(X,E)/kT} = e^{-U(X)/kT}\left(1 + \frac{3\epsilon_s}{2\epsilon_s+1}\frac{M(X)E}{kT}\cos\theta + \ldots\right). \qquad (7.7)$$

Now
$$\int M(X)\cos\theta\, e^{-U(X)/kT}\,dX = 0 \qquad (7.8)$$

because this integral is proportional to the average moment in the **E**-direction in the absence of a macroscopic field. Therefore, inserting from 7.7 into 7.3, making use of 7.8,

$$M_E = \frac{3\epsilon_s}{2\epsilon_s+1}\frac{EJ}{kT}\int M^2(X)\cos^2\theta\, e^{-U(X)/kT}\,dX, \qquad (7.9)$$

where
$$1/J = \int e^{-U(X)/kT}\,dX. \qquad (7.10)$$

Assume now that the field **E** may have any direction relative to a given $\mathbf{M}(X)$, and average over all these directions. Then $\cos^2\theta$ will be replaced by its average value, i.e. by $\tfrac{1}{3}$. Inserting this value into equation 7.9, and making use of 1.9, yields

$$\epsilon_s - 1 = \frac{4\pi}{3V} \frac{3\epsilon_s}{2\epsilon_s+1} \frac{\overline{M^2}}{kT}, \qquad (7.11)$$

where
$$\overline{M^2} = J \int M^2(X) e^{-U(X)/kT} dX \qquad (7.12)$$

is the average value of $M^2(X)$ in the absence of a macroscopic field. Equation 7.11 represents our first general result. It shows that the dielectric constant can be expressed in terms of $\overline{M^2}$, the mean square of the spontaneous polarization of a sphere of the dielectric embedded in a large specimen of the same material.

Before developing the general theory further, equation 7.11 will be shown to be self-consistent. This means that 7.11 must be fulfilled identically if the spherical region is treated on a macroscopic basis. For this purpose the following theorem will be employed (cf. reference *T1*). Suppose the free energy $F(\alpha_1, \alpha_2,...)$ of a system depends on a number of macroscopic parameters $\alpha_1, \alpha_2,...$. Then the probability of finding the system in a range $d\alpha_1 d\alpha_2...$ is given by

$$e^{-F(\alpha_1,\alpha_2...)/kT} d\alpha_1 d\alpha_2... \Big/ \int e^{-F(\alpha_1,\alpha_2...)/kT} d\alpha_1 d\alpha_2....$$

Also it will be noted that the quantity $(d\alpha_1 d\alpha_2...)$ in the above expression represents the volume of the 'α-space' element between $(\alpha_1, \alpha_2...)$ and $(\alpha_1+d\alpha_1, \alpha_2+d\alpha_2,...)$. In the present case, since the free energy depends on the absolute magnitude of M only, this element is the shell between the two spheres of radius M and $M+dM$ in the M-space, and has the volume $4\pi M^2 dM$. Since M may in theory have any value between zero and infinity, the probability that the spherical region will have a moment between M and $M+dM$ is

$$e^{-F(M)/kT} M^2 dM \Big/ \int_0^\infty e^{-F(M)/kT} M^2 dM. \qquad (7.13)$$

Hence $\overline{M^2} = \int_0^\infty M^2 e^{-F(M)/kT} M^2 dM \Big/ \int_0^\infty e^{-F(M)/kT} M^2 dM. \qquad (7.14)$

Comparing this with equation 7.12, and remembering that according to 3.3 the entropy S is given by $S = (U-F)/T$, it follows that
$$dX \sim e^{S/k} M^2 \, dM \tag{7.15}$$
represents the number of states of the spherical region, in a range between M and $M+dM$. This equation might have been used as an alternative starting-point for the macroscopic considerations based on 7.12.

Following the procedure used in the case of the energy $U(X)$, the free energy $F(M)$ can be considered as composed of an internal free energy $F_i(M)$, and an external one, $F_e(M)$, due to the interaction with the surroundings. The former is the self-energy F_s calculated in the appendix (A 2.37),
$$F_i(M) = \frac{2\pi M^2}{3V} \frac{\epsilon_s + 2}{\epsilon_s - 1}; \tag{7.16}$$
the latter can be obtained with the help of the reaction-field **R** (cf. equation 5.10). It is composed of the energy $-\mathbf{MR}$ of the dipole **M** in the field **R**, and of the free energy required to polarize the surroundings accordingly. This latter should be proportional to R^2, say βR^2. Then
$$F_e(M) = -\mathbf{MR} + \beta R^2. \tag{7.17}$$
To find β, F_e can be considered to depend on the parameters R_x, R_y, R_z [for a fixed value of M], and hence since for a system in equilibrium the free energy should be a minimum
$$\frac{\partial F_e}{\partial R_x} = 0, \qquad \frac{\partial F_e}{\partial R_y} = 0, \qquad \frac{\partial F_e}{\partial R_z} = 0$$
or
$$2\beta \mathbf{R} = \mathbf{M}.$$
Inserting this into 7.17, using 5.10,
$$F_e(M) = -\mathbf{MR} + \tfrac{1}{2}\mathbf{MR} = -\tfrac{1}{2}\mathbf{MR} = -\frac{4\pi M^2}{3V} \frac{\epsilon_s - 1}{2\epsilon_s + 1}. \tag{7.18}$$
Thus from 7.16 and 7.18
$$F = F_i(M) + F_e(M) = \frac{2\pi M^2}{V} \frac{3\epsilon_s}{(2\epsilon_s + 1)(\epsilon_s - 1)}, \tag{7.19}$$

and hence with† 7.14,

$$\overline{M^2} = \tfrac{3}{2}kT\frac{V}{2\pi}\frac{(2\epsilon_s+1)(\epsilon_s-1)}{3\epsilon_s}, \qquad (7.20)$$

which shows that equation 7.11 actually is fulfilled identically. The derivation of this result demonstrates the importance of the long-range interaction terms—included in $F_e(M)$—in equation 7.11. This means that in view of this interaction the fluctuations $\overline{M^2}$ of a sphere embedded in its own medium are different from the corresponding fluctuations for a sphere in vacuum. In the latter case $F = F_i$, and 7.14 leads to

$$\overline{M^2_{\text{vac}}} = \tfrac{3}{2}kT\frac{3V}{2\pi}\frac{\epsilon_s-1}{\epsilon_s+2}. \qquad (7.21)$$

Comparing 7.20 with 7.21 shows that $\overline{M^2}$ becomes relatively large for substances with high dielectric constant in contrast to $\overline{M^2_{\text{vac}}}$ which does not depend appreciably on ϵ_s if $\epsilon_s \gg 1$.

In returning now to the further development of equation 7.11 a method will be used which forms a generalization of the method employed by Kirkwood (*K4*) in the case of liquids consisting of rigid dipoles. Assume that the spherical region is composed of N molecules or other groups of atoms in such a way that each group has the same average polarization in an external field. In a pure liquid, for instance, each group contains one molecule; in a crystal it contains all the particles of a unit cell. The spherical region can then be divided into N 'units' each of which makes the same contribution to M_E. In view of equations 7.9–7.12 it follows that $\overline{M^2}$ is also composed of N equal terms. This finds its mathematical expression in equations 7.31 and 7.32. A more direct proof that all these terms are equal is given below after equation 7.30. Each such unit contains the same number, say k, of elementary charges which in each case can be arranged in a similar way relative to each other, and relative to their surroundings. Let $\mathbf{r}_{j1}, \mathbf{r}_{j2},\ldots \mathbf{r}_{jk}$ denote the displacements of the k elementary charges of the jth unit, and let

$$x_j = (\mathbf{r}_{j1}, \mathbf{r}_{j2},\ldots, \mathbf{r}_{jk})$$

† Using $\int_0^\infty x^2 e^{-x^2}x^2\,dx = \tfrac{3}{2}\int_0^\infty e^{-x^2}x^2\,dx.$

denote the whole set of displacements of the jth unit. Then the set of displacements X of all the charges in the whole spherical region is composed of the sets of displacements $x_1, x_2, ..., x_j, ..., x_N$ of all the N units. Correspondingly the volume element dX (cf. 7.21) is the product of the volume elements

$$dx_j = d\mathbf{r}_{j1} d\mathbf{r}_{j2} ... d\mathbf{r}_{jk}$$

of all the units

$$dX = dx_1 dx_2 ... dx_j ... dx_N.$$

This means that in an integration over the displacements of all the elementary charges of the spherical region one can start integrating over the displacements of the charges in the first unit, denoted by dx_1, and so on. Alternatively, the volume element dX may be written as

$$dX = dX_j dx_j, \tag{7.22}$$

where $$dX_j = dx_1 dx_2 ... dx_{j-1} dx_{j+1} ... dx_N \tag{7.23}$$

refers to integration over the whole spherical region except the jth unit.

Now let $\mathbf{m}(x_j)$ be the electric dipole moment of the jth unit. Then in view of 4.7 the moment $\mathbf{M}(X)$ of the whole region is composed as the vector sum of all the $\mathbf{m}(x_j)$,

$$\mathbf{M}(X) = \sum_{j=1}^{N} \mathbf{m}(x_j). \tag{7.24}$$

Therefore $$M^2(X) = \mathbf{M}(X)\mathbf{M}(X) = \sum_{j=1}^{N} \mathbf{m}(x_j)\mathbf{M}(X). \tag{7.25}$$

Inserting this into 7.12 leads to

$$\overline{M^2} = \sum_{j=1}^{N} J \int \mathbf{m}(x_j)\mathbf{M}(X) e^{-U(X)/kT} dX. \tag{7.26}$$

The integrations in each term of this sum will now be carried out in two steps, as indicated by equation 7.22. In the jth term the integration will first be carried out over the whole spherical region except the jth unit, and subsequently over this unit. For the first step (volume element dX_j) the quantity $\mathbf{m}(x_j)$ must be treated as a constant because it depends only on the displacements x_j of the jth unit. The term of 7.26 corresponding to this unit can be written as

$$J \int \mathbf{m}(x_j)\mathbf{M}(X)e^{-U(X)/kT}dX$$
$$= J \int \mathbf{m}(x_j)\left(\int \mathbf{M}(X)e^{-U(X)/kT}dX_j\right)dx_j$$
$$= \int \mathbf{m}(x_j)\mathbf{m}^*(x_j)p(x_j)dx_j, \quad (7.27)$$

where
$$\mathbf{m}^*(x_j) = \int \mathbf{M}(X)e^{-U(X)/kT}dX_j \Big/ \int e^{-U(X)/kT}dX_j, \quad (7.28)$$
and (cf. 7.10)
$$p(x_j) = \int e^{-U(X)/kT}dX_j \Big/ \int e^{-U(X)/kT}dX \quad (7.29)$$

have been introduced. $\mathbf{m}^*(x_j)$ represents the average moment of the whole sphere for a fixed set x_j of displacements of the jth unit leading to a moment $\mathbf{m}(x_j)$. $p(x_j)$ is the probability of finding the jth unit with this particular set of displacements. Inserting from 7.27 into 7.26 yields

$$\overline{M^2} = \sum_{j=1}^{N} \int \mathbf{m}(x_j)\mathbf{m}^*(x_j)p(x_j)\,dx_j. \quad (7.30)$$

Practically all terms of this sum are equal. For $\mathbf{m}^*$ represents the moment of a large spherical region polarized by one (say the jth) of its units whose moment is kept at a value $\mathbf{m}$. According to electrostatics (cf. Appendix A 2.ii) the same moment $\mathbf{m}^*$ is contained in any other sphere enclosing the jth unit (not necessarily co-centric), but with no further restriction of its position and radius so long as it is sufficiently large to be treated macroscopically. The actual value of $\mathbf{m}^*$ is thus determined by short-range interaction and is independent of the position of the jth unit as long as its distance from the surface is sufficiently large to allow its interaction with the outside to be treated on a macroscopic basis. The number of units for which this does not hold can be made very small compared with the total number N if the spherical region is sufficiently large.

The equivalence of all N terms in the sum 7.30 means, of course, that all units have the same average polarization. This fact might indeed have been used to show the equivalence of all the above terms in a much simpler way. The above considerations give, however, some insight into the forces determining **m***. It follows from these considerations that a region makes a contribution to **m*** only if the average moment induced in it by the jth unit cannot be obtained by treating the jth unit as either a point dipole or as a polarized sphere with moment **m**. Thus it is essentially the short-range forces and the deviation of the shape of molecules from a sphere which makes **m*** different from **m**.

It now follows that

$$\overline{\mathbf{mm^*}} = \int \mathbf{m}(x_1)\mathbf{m^*}(x_1)p(x_1)\,dx_1$$
$$= \ldots = \int \mathbf{m}(x_j)\mathbf{m^*}(x_j)p(x_j)\,dx_j = \ldots \quad (7.31)$$

and hence with 7.30,

$$\overline{M^2} = N\overline{\mathbf{mm^*}}. \quad (7.32)$$

Inserting this into equation 7.11 leads to our final expression for the static dielectric constant,

$$\epsilon_s - 1 = \frac{3\epsilon_s}{2\epsilon_s+1} \frac{4\pi N_0}{3} \frac{\overline{\mathbf{mm^*}}}{kT}, \quad (7.33)$$

where
$$N_0 = N/V \quad (7.34)$$

is the number of units per unit volume. By its definition 7.28 **m*** represents the average dipole moment of a spherical region embedded in its own medium, if one of its units is kept in a given configuration leading to a dipole moment **m**. $\overline{\mathbf{mm^*}}$ is the average value of the product **mm*** taking into account all possible configurations and weighing them according to the probability of finding the unit in such a configuration. **m*** differs from **m** because of the existence of short-range forces or because of the non-spherical shape of molecules.

Equation 7.33 is perfectly general. It will now be specialized by separating the contribution to ϵ_s due to elastic displacements of electrons by treating them on a macroscopic basis. This contribution can be measured with the help of the optical refractive index n because at optical frequencies all other contributions have ceased to exist in view of the higher inertia of the heavier particles. If $\mathbf{M}_{el}$ is the electric moment due to electronic displacement by a field $\mathbf{E}$, on the assumption that atomic nuclei are not displaced, then

$$n^2 - 1 = 4\pi M_{el}/VE. \qquad (7.35)$$

Since according to Maxwell's law the dielectric constant at high frequencies is equal to the square of the refractive index, this relation follows directly from equation 1.9.

The spherical region now consists of charges e_i which follow the laws of statistical mechanics and which are embedded in a continuous medium with dielectric constant n^2. Thus if M_E is the moment due to all other displacements, $M_E + M_{el}$ is the total moment of the substance. Hence equation 1.9 becomes

$$\epsilon_s - 1 = 4\pi(M_E + M_{el})/VE,$$

or using 7.35 $\qquad \epsilon_s - n^2 = 4\pi M_E/VE. \qquad (7.36)$

Instead of calculating the electronic polarization we may introduce it through 7.36 as an empirical quantity on the understanding that n is the optical refractive index. Nearly all of the above developments remain unaltered if $\mathbf{M}(X)$, $\mathbf{m}$, and $\mathbf{m}^*$ now refer to non-electronic displacements. Alterations have to be introduced only at the following two points:

(i) In equation 7.5 the cavity field $\mathbf{G}$ must now be replaced by $\mathbf{G}'$, the field inside a spherical cavity with dielectric constant n^2 instead of in an empty cavity. Then according to Appendix A 2.14,

$$\mathbf{G}' = \frac{3\epsilon_s}{2\epsilon_s + n^2}\mathbf{E}, \qquad (7.37)$$

i.e. the denominator $(2\epsilon_s + 1)$ in 7.5 and in the following equations must now be replaced by $(2\epsilon_s + n^2)$.

(ii) In deriving 7.11 from 7.9 equation 7.36 should be used instead of 1.9. Hence 7.11 will be replaced by

$$\epsilon_s - n^2 = \frac{4\pi}{3V} \frac{3\epsilon_s}{2\epsilon_s + n^2} \frac{\overline{M^2}}{kT}. \qquad (7.38)$$

All further developments remain unaltered, leading to

$$\epsilon_s - n^2 = \frac{3\epsilon_s}{2\epsilon_s + n^2} \frac{4\pi N_0}{3} \frac{\overline{\mathbf{mm}^*}}{kT} \qquad (7.39)$$

instead of equation 7.33. It should be realized that in considering the interaction energy $U(X)$ required for the calculation of **m** and **m*** (cf. equations 7.27–7.31) the electronic polarization has to be included on a macroscopic basis.

Finally, it should be noted that the difference between 7.39 and 7.33 is of rather a trivial nature whenever $\epsilon_s \gg n^2$, i.e. when the electronic contributions are small. In this case

$$\frac{3\epsilon_s}{2\epsilon_s+n^2} \simeq \frac{3\epsilon_s}{2\epsilon_s+1} \simeq \frac{3}{2}, \qquad \epsilon_s \gg n^2 \qquad (7.40)$$

and 7.39 becomes

$$\epsilon_s - n^2 = 2\pi N_0 \frac{\overline{\mathbf{mm^*}}}{kT}, \qquad \epsilon_s \gg n^2 \qquad (7.41)$$

while 7.33 would have $\epsilon_s - 1$ on the left-hand side, which thus differs from 7.41 by the optical contributions $n^2 - 1$.

Mixtures

Some substances contain a number of distinctly different groups of charges such as the different types of molecules in a mixture, or the positive and negative ions in ionic crystals. In these cases a unit cell would often be uncomfortably large or have an undesirable shape. Whenever these groups are well separated from each other it is possible to derive a formula for ϵ_s in which the contributions of these different groups are separated. Suppose that there are z different types of these groups. Then z different kinds of units will exist, each being representative for one type. If the substance contains per c.c. $N_1, N_2, ..., N_s, ..., N_z$ such units of the 1st, 2nd,...zth kind, then 7.24 can be written as

$$\frac{\mathbf{M}(X)}{V} = \sum_{j=1}^{N_1} \mathbf{m}(x_j) + \sum_{j=N_1+1}^{N_1+N_2} \mathbf{m}(x_j) + ... \qquad (7.42)$$

For the whole further development each group can be treated separately, so that instead of 7.33 the final result is

$$\epsilon_s - 1 = \frac{3\epsilon_s}{2\epsilon_s+1} \frac{4\pi}{3kT} \sum_{s=1}^{z} \overline{\mathbf{m}_s \mathbf{m}_s^*} N_s, \qquad (7.43)$$

where $\mathbf{m}_s^*$ is the average dipole moment of a spherical region of the substance embedded in its own medium if one unit of the sth kind is kept in a configuration corresponding to a moment $\mathbf{m}_s$. $\overline{\mathbf{m}_s \mathbf{m}_s^*}$ is the average value of $\mathbf{m}_s \mathbf{m}_s^*$. As before, the spherical region must be large compared with a region just big enough to have the same dielectric properties as a macroscopic specimen.

Summary

A short summary of the results of § 7 is given below for the benefit of the reader who has not studied it in detail.

(i) The static dielectric constant ϵ_s according to 7.11 and 7.21 satisfies exactly the relations

$$\epsilon_s - 1 = \frac{4\pi}{3V} \frac{3\epsilon_s}{2\epsilon_s+1} \frac{\overline{M^2}}{kT} = \frac{4\pi}{3V} \frac{\epsilon_s+2}{3} \frac{\overline{M^2_{\text{vac}}}}{kT}, \qquad (7.44)$$

where $\overline{M^2}$ is the mean square of the spontaneous dipole moment of a sufficiently large sphere of dielectric material of volume V embedded in its own medium; $\overline{M^2_{\text{vac}}}$ is the corresponding quantity for a sphere in vacuum.

(ii) If the sphere consists of components all of which on an average make the same contribution to the polarization, equation 7.44 can be developed further into equation 7.33 which is also of a very general nature.

By treating the electronic contribution on a macroscopic basis the less general equation 7.39 follows.

8. Special cases

The results of the previous section are contained in formula 7.33 [also 7.11] for the dielectric constant ϵ_s which is valid in a very general way, and in the more specialized equations 7.39 and 7.43. All these formulae represent ϵ_s in terms of quantities $\overline{\mathbf{mm}^*}$ or $\overline{M^2}$ which refer to properties of the material in the absence of a field. To calculate these quantities requires a detailed knowledge of the structure of the substance in question and of the interaction between the particles of which it is composed. In general such a calculation cannot be carried out without the use of approximations. The importance of the general formulae lies in the possibility of deriving from it approximate formulae of various types, each with a clearly defined range of validity. It also shows that even for a restricted class of substances, such as dipolar liquids, one cannot hope to find a simple formula representing ϵ_s in terms of a few parameters (e.g. the Onsager formula), which is valid over the whole stable range of these substances. Thus it will be found, for instance, that for dipolar liquids the Onsager formula should hold asymptotically at high temperatures only, though the deviations from it at lower temperatures may be very different for different substances.

As a first step to get acquainted with the handling of the above-mentioned formulae one may assume that all short-range forces can be neglected. For spherical molecules this should according to § 6 lead to either the Clausius–Mossotti or the Onsager formula, depending on whether one considers non-polar or polar molecules. The calculations proving this are carried out in the appendix (A 3). They demonstrate, as the main difference between these two cases, that for elastic displacement (non-polar molecules) the energy of a unit depends on its dipole moment m; hence it is shown that $\overline{m^2}$ (which for negligible short-range interaction is equal to $\overline{mm^*}$) increases proportional to the absolute temperature T, and the dielectric constant becomes independent of the temperature. For rigid dipoles, to take the other extreme, the moment of a unit is equal to the dipole moment μ, and hence $\overline{m^2} = \mu^2$ independent of temperature. Polar molecules, of course, show always both dipolar and elastic contributions to ϵ_s.

Polar liquids; Kirkwood's formula

We shall now proceed to consider the general case of polar liquids consisting of molecules with an intrinsic dipole moment and with a polarizability α. Liquids of this type have been already investigated in § 6 on the assumption that short-range interaction can be neglected; this led to the Onsager formula 6.36. This assumption will no longer be made in the present calculation.

It was found in § 6 that a molecule in the liquid state has a dipole moment which is different from that in the gaseous state (μ_v). For spherical molecules the difference is due to the polarization of the molecule by the reaction field of the surroundings.

For no other case than for spherical molecules is it possible to calculate the ratio of the moments in any simple precise manner. In view of this difficulty it seems reasonable to treat the effect of the polarization of molecules in a macroscopic way, assuming (as in § 6) that the main contributions are due to electron displacement. The liquid will thus be considered as consisting of a continuous medium with dielectric constant n^2 (n = optical refractive index) containing dipoles with a moment μ. In this model a spherical molecule in vacuum consists of a sphere of the continuous medium containing a dipole μ at its centre. According to the appendix (A 2.32) the moment of such a molecule is

$$\mu_v = \frac{3}{n^2+2}\mu, \quad \text{spherical molecules.} \tag{8.1}$$

For spherical molecules the present model permits us, therefore, to express the moment μ of the dipoles by the moment μ_v of a free molecule.

On the basis of our model the general formula for the dielectric constant is given by equation 7.39. A unit contains just one dipole μ and its moment is, therefore,

$$\mathbf{m} = \boldsymbol{\mu}. \tag{8.2}$$

Since all contributions to $\mathbf{m}^*$ are due to dipole orientation we introduce $\boldsymbol{\mu}^*$ by

$$\mathbf{m}^* = \boldsymbol{\mu}^* \tag{8.3}$$

as the average moment due to the dipoles of a spherical region if one of its dipoles is kept in a fixed direction. Equation 7.39 contains the quantity $\overline{\mathbf{mm}^*}$, where the bar indicates averaging over all possible values of $\mathbf{m}$. In the present model the only variable is the direction of the dipole μ. Since in a liquid all dipolar directions are equivalent, $\overline{\boldsymbol{\mu\mu}^*}$ will have the same value for all these directions. Thus

$$\overline{\mathbf{mm}^*} = \overline{\boldsymbol{\mu\mu}^*} = \boldsymbol{\mu\mu}^*. \tag{8.4}$$

Inserting 8.4 into 7.39 leads to the Kirkwood [K4] formula

$$\epsilon_s - n^2 = \frac{3\epsilon_s}{2\epsilon_s + n^2} \frac{4\pi N_0 \boldsymbol{\mu\mu}^*}{3kT}. \tag{8.5}$$

In order to make use of this formula we must calculate $\boldsymbol{\mu}^*$.

To simplify this calculation it seems permissible to assume that short-range interaction between nearest neighbours only need be considered (cf. first section of § 5). In this case $\boldsymbol{\mu}^*$ is the vector sum of the moment $\boldsymbol{\mu}$ of the central dipole, kept in a fixed direction, and of the average of the sum of the moments of the nearest neighbours. Therefore if z is the average number of nearest neighbours,

$$\boldsymbol{\mu\mu}^* = \mu^2(1+z\overline{\cos\gamma}), \tag{8.6}$$

where $\overline{\cos\gamma}$ is the average of the cosine of the angle between neighbouring dipoles. This average value has to be calculated by applying equation 7.28 for $\mathbf{m}^*$ to the present case. Since the directions of the dipoles are the only variables, this equation reduces to

$$\overline{\cos\gamma} = \int \cos\gamma\, e^{-U/kT}\, d\omega_1\, d\omega_2 \Big/ \int e^{-U/kT}\, d\omega_1\, d\omega_2, \tag{8.7}$$

where U is the part of the energy of interaction between neighbouring molecules in the liquid which depends on the angle between their dipoles; this energy may depend on the state of other molecules of the liquid, and U is then the energy averaged over all states of the other molecules (considering their probabilities). $d\omega_1$ and $d\omega_2$ are the surface elements of the solid angles of the directions of the two dipoles.

The actual value of $\overline{\cos\gamma}$ depends on details of the interaction between the two molecules which may be very different from the electrostatic interaction between point dipoles as pointed out in § 5; repulsion forces, chemical bond, and other types of interaction must be considered. It should be noted, however, that a large value of U does not necessarily lead to large values of $\overline{\cos\gamma}$; to do so U must also have the correct symmetry. Thus if, for instance, U is proportional to an even power of $\cos\gamma$ $e^{-U/kT}$ is also an even function of $\cos\gamma$, and $\overline{\cos\gamma}$ therefore vanishes, in contrast to the case in which U is an odd function of $\cos\gamma$. This means that an interaction tending to direct dipoles with equal probability either parallel or antiparallel does not play any role in the determination of $\boldsymbol{\mu\mu}^*$. Either of the two types of interaction leads, however, to a restriction of free rotation of a molecule termed by Debye [*D3*] as hindered rotation. Hindered rotation, as we see, does not necessarily have an influence on the dielectric constant.

The energy U need not only depend on the angle between the dipoles, but may also be a function of the direction of the vector joining the dipoles. Even so, we can always put

$$U = U_{\text{even}} + U_{\text{odd}}, \tag{8.8}$$

where reversal of the direction of one of the dipoles does not alter U_{even} but changes the sign of U_{odd}.

Now let us consider that the temperature is sufficiently high to make

$$kT \gg |U_{\text{odd}}| \tag{8.9}$$

for all values of U_{odd}. Then

$$e^{-U/kT} \simeq e^{-U_{\text{even}}/kT}\left(1 - \frac{U_{\text{odd}}}{kT}\right), \tag{8.10}$$

and hence, using

$$\int \cos\gamma\, e^{-U_{\text{even}}/kT}\, d\omega_1\, d\omega_2 = 0,$$

one finds

$$\overline{\cos\gamma} \simeq -\frac{U_0}{kT}, \tag{8.11}$$

where

$$U_0 = \int e^{-U_{\text{even}}/kT} U_{\text{odd}} \cos\gamma\, d\omega_1\, d\omega_2 \Big/ \int e^{-U_{\text{even}}/kT} d\omega_1\, d\omega_2 \tag{8.12}$$

since

$$\int e^{-U_{\text{even}}/kT} U_{\text{odd}}\, d\omega_1\, d\omega_2 = 0.$$

The energy U_0 may be positive or negative depending on whether the interaction tends to orient neighbouring dipoles anti-parallel or parallel.

Inserting 8.6 into 8.5, we find a specialized form of the Kirkwood formula,

$$\epsilon_s - n^2 = \frac{3\epsilon_s}{2\epsilon_s + n^2} \frac{4\pi N_0 \mu^2}{3kT}(1 + z\overline{\cos\gamma}). \tag{8.13}$$

In the particular case of spherical molecules μ can be expressed by the vacuum moment μ_v with the help of 8.1, and hence

$$\epsilon_s - n^2 = \frac{3\epsilon_s}{2\epsilon_s + n^2}\left(\frac{n^2+2}{3}\right)^2 \frac{4\pi N_0 \mu_v^2}{3kT}(1 + z\overline{\cos\gamma})$$

for spherical molecules. (8.14)

This equation differs from the Onsager formula 6.36 by the

$z\overline{\cos\gamma}$ term. According to 8.11 this term tends towards zero at temperatures at which $kT \gg |U_0|$. Therefore it is found that the Onsager formula should hold in liquids at temperatures for which kT is large compared with the directional part U_{odd} of the interaction energy. This energy may have different orders of magnitude for different liquids; and this may lead to very different ranges of validity of the Onsager formula. In some liquids this range may fall into the whole-liquid range at normal

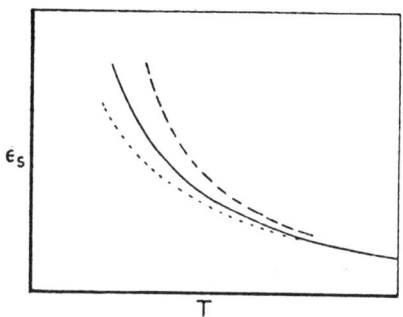

FIG. 5. Temperature-dependence of the dielectric constant ϵ_s in the high-temperature region. The full line represents Onsager's formula. Short-range interaction tending to orient dipoles parallel ($\overline{\cos\gamma} > 0$) leads to larger values of ϵ_s (-----); the opposite case ($\overline{\cos\gamma} < 0$) leads to smaller values (·····).

pressures, whereas for others $z\overline{\cos\gamma}$ may be appreciable in the normal range of the liquid and may have either sign (cf. Fig. 5) thus making the Onsager formula inapplicable in these cases. In gases, in view of the large distance between molecules, dipolar forces only need be considered. For very small densities interaction may be completely neglected, leading to equation 6.15 for ϵ_s. For higher densities Onsager's formula should hold. This has been shown more directly by van Vleck [V2].

Dipolar solids

The average potential energy of a polar molecule in a crystalline solid in general depends on the direction of its dipole relative to the crystal axes. In liquids in contrast, though the dipoles of neighbouring molecules have a tendency to orient themselves in definite directions relative to each other, the average energy of a single dipole is the same in all directions because a liquid does not possess any preferential direction. The average potential energy of a molecule in a solid is often described as being due to a crystalline field acting on the dipoles.† The crystalline field, being due to the interaction between molecules, normally depends on temperature. Usually there are several dipole directions for which the average energy of a molecule has a relative minimum; they will be called equilibrium directions. The potential barrier between these equilibrium directions is in general very high and prevents free rotation of the molecules even at temperatures near the melting-point. The hypothesis that above a critical temperature molecules should be able to rotate freely was introduced by Pauling [*P1*] to explain the sudden change of the dielectric properties of many solids at a critical temperature. This idea was a useful working hypothesis, but the present evidence seems to indicate that rotation does not occur in most solids (e.g. *L1*). Instead, the transition has to be considered as leading from an ordered arrangement of dipoles to disorder.

This transition from order into disorder will be used as a guiding principle in discussing the general behaviour of the dielectric constant ϵ_s of polar solids. Its detailed properties, of course, depend on details of the crystalline field, but the main features can be discussed without such a detailed knowledge.‡ To describe the main features of an order-disorder transition consider, a simple two-dimensional model consisting of dipoles arranged in a cubic face centred lattice. The crystalline field is assumed to lead for each molecule to two equilibrium positions with opposite dipole direction. At the absolute zero of temperature the solid will be in its lowest energy state. In this state the dipoles are arranged in an ordered way, but there are a number of possibilities for doing this. The energy of these ordered states depends on the particular interaction between the molecules which, as has been pointed out before, contains interactions of various types (e.g. dipolar, repulsive, etc.). These ordered arrangements can be classified according to whether or not they

† The effect of various types of crystalline fields on the average dipole direction was considered by Bauer [*B1*], Frank [*F3*], and others.
‡ Various calculations of the behaviour of the dielectric constant in order-disorder transitions have been carried out for special models [*K5*, *F8*].

lead to a residual dipole moment of the crystal. Thus, if all dipoles are parallel (Fig. 6 a), the solid has a permanent dipole moment, but if, for instance, dipoles in the corners have the opposite direction from those at the centres (Fig. 6 b) the dipole moment of the crystal vanishes. In the former case the energy

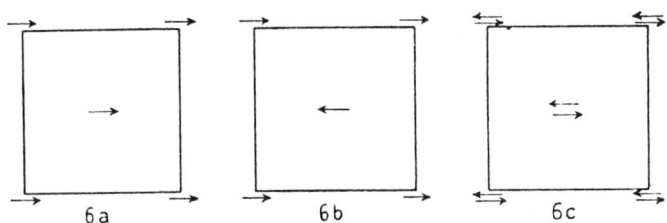

FIG. 6. Equilibrium positions of dipoles for a simple model of a crystalline solid. Ordered state: (a) in the case of permanent polarization; (b) for vanishing polarization (two dipoles per unit cell); (c) disordered state, both directions have equal probabilities.

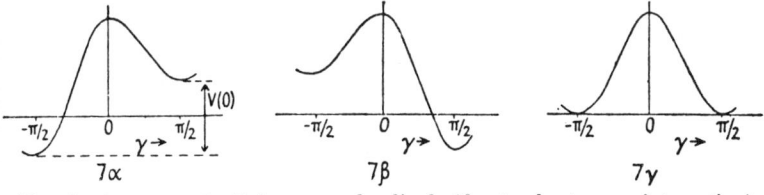

FIG. 7. Average potential energy of a dipole (due to short-range interaction) as a function of its direction for the model of Fig. 6, assuming that all the other dipoles stay in the fixed directions indicated in Fig. 6: (α) for the central dipole of Fig. 6 b; (β) for a corner dipole of Fig. 6 b or for any dipole of Fig. 6 a; (γ) for the disordered case, Fig. 6 c.

of a dipole is lower when it is directed to the right than for the opposite directions (cf. Fig. 7 β). In the latter case the energy of the corner-dipoles behaves in a similar way, but for the centre-dipoles the two directions are exchanged (Fig. 7 α). The internal fields for these two types of dipoles are, therefore, different; they transform into each other by a reflection on a plane perpendicular to the dipole direction. In the former case (a) all lattice-points are equivalent and a unit cell contains one single dipole. In the latter case (b) two different types of sites, corners and centres, have to be distinguished, and a unit cell contains two dipoles. Alternatively, the crystal may be considered as composed of two lattices of type (a), but with opposite directions of the dipoles.

Clearly the division of ordered states according to whether or not they lead to a residual moment holds quite generally (i.e. for three-dimensional lattices as well), although there may be structures leading to more than two equilibrium directions for each dipole. In the following the case of two equilibrium directions only will be considered; most results hold qualitatively, however, for more complicated structures as well.

Now, starting from an ordered state at the absolute zero of temperature, imagine the temperature to increase gradually. This will cause some dipoles to turn into the second equilibrium position in which they have a higher energy, thus creating some disorder. This in turn will cause a decrease of the average energy difference between the two directions because the energy of interaction has its lowest value in the completely ordered state. The average energy difference of a molecule in the two equilibrium directions is thus a function of temperature, say

$$V(T) \geqslant 0,$$

which decreases with increasing temperature. It is found that a critical temperature T_0 exists at which it vanishes (Fig. 7 γ).

Let us now define the direction which a dipole at a given lattice-point has in the completely ordered state (at $T = 0$) as the 'right' direction, and the opposite direction as the 'wrong' direction. Thus in Fig. 6 a, → is the right direction; in Fig. 6 b, → is the right direction for corner-dipoles and ← is the right direction for centre-dipoles. Let w be the probability of finding a dipole in the wrong direction, and $(1-w)$ therefore, the probability of finding it in the right direction. Since $V(T)$ is the energy difference between the two positions, it follows from statistical mechanics that

$$\frac{w}{1-w} = e^{-V(T)/kT}, \tag{8.15}$$

and hence

$$w = \frac{e^{-V(T)/kT}}{1+e^{-V(T)/kT}}, \qquad 1-w = \frac{1}{1+e^{-V(T)/kT}}. \tag{8.16}$$

A calculation of the temperature dependence of $V(T)$, and hence of w is very difficult, but approximations have been devised to simplify it. A summary of these methods can be found in reference [N1]. We shall not give details of such calculations

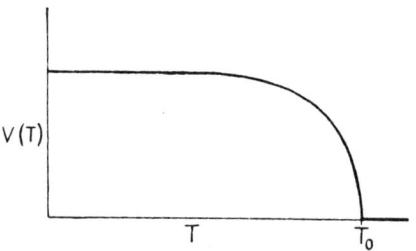

Fig. 8. Temperature-dependence of the average energy difference $V(T)$ between opposite dipole directions, schematically.

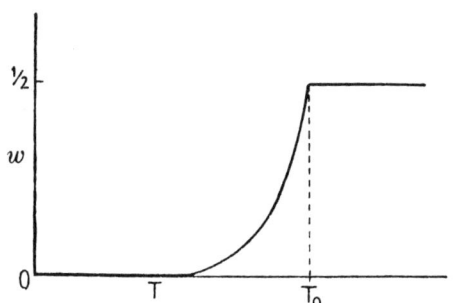

Fig. 9. Temperature-dependence of the probability $w(T)$ of finding the dipole in the 'wrong' direction, schematically.

here. It is sufficient for us to note (cf. Fig. 8) that near $T = 0$, $V(T)$ is nearly constant, but as T approaches T_0 it decreases rapidly towards zero, which it reaches at $T = T_0$. Thus

$$V(T) = 0 \text{ if } T \geqslant T_0.$$

It follows (cf. Fig. 9) that for low temperatures

$$w \simeq e^{-V(0)/kT} \ll 1 \quad \text{if } kT \ll V(0), \tag{8.17}$$

whereas $\qquad w = \tfrac{1}{2} \quad \text{if } T \geqslant T_0. \tag{8.18}$

Thus above T_0 both dipole directions have equal probabilities, i.e. the lattice is disordered (cf. Fig. 6 c). Calculations also show that T_0 is of the order of $V(0)/k$; the exact value depends on details of the structure.

The quantity $(1-2w)$ is a measure for the degree of order of the lattice because $1-2w = 1$ at $T = 0$, and $1-2w = 0$ if $T \geqslant T_0$. This type of order is often called long-distance order because it defines right and wrong directions for any lattice-point. In contrast, short-distance order is the order of neighbours relative to each other. It means that, in view of the interaction, the direction of a dipole is always influenced by the directions of its neighbours. Each dipole tends to orient itself in a certain direction relative to its neighbours. Long-distance order vanishes in the disordered state, or in liquids. Short-distance order persists, however, though it decreases with increasing temperature. In fact the absolute value of $\overline{\cos \gamma}$ introduced above for liquids is a measure for the short-distance order.

Let us now discuss the dielectric properties of ordered solids near $T = 0$. If the order is of type (a), the solid is permanently polarized. This case will not be further considered here. For the other kind of order the only contribution of the dipoles to the dielectric constant is due to elastic displacement of the dipole direction by an external field. Thus near $T = 0$ the dielectric constant is larger than n^2, say $\epsilon_s = \epsilon_\infty$, where n is the optical refractive index due to elastic displacement of electrons (cf. §§ 6 and 7). The difference $\epsilon_\infty - n^2$ is usually small.

At higher temperatures, where dipoles are capable of changing their directions, ϵ_s increases with temperature, as will be shown presently. As in the case of liquids, contributions due to elastic displacements will be treated in a macroscopic way, and the following considerations are, therefore, based on equation 7.39. For the type of order which we consider at present a unit cell contains two dipoles as mentioned before, and at $T = 0$ these dipoles have opposite direction. For the two-dimensional case discussed above the unit cell contains a corner- and a centre-dipole as shown in Fig. 6 b. The moment **m** of the unit cell depends on the directions of the dipoles. Since the potential barriers between the equilibrium directions are very high, it

SPECIAL CASES

seems sufficient to consider equilibrium directions only. The moment **m** can thus have a discrete number of values $\mathbf{m}_1$, $\mathbf{m}_2$,.... If p_i is the probability that the unit cell has the moment $\mathbf{m}_i$, equation 7.31 applied to the present case becomes

$$\overline{\mathbf{mm}^*} = \sum \mathbf{m}_i \mathbf{m}_i^* \, p_i, \qquad (8.19)$$

where the sum extends over all values of $\mathbf{m}_i$. $\mathbf{m}_i^*$ is the average moment of a sufficiently large spherical region around the unit cell if the moment of the latter is kept at $\mathbf{m}_i$.

In our particular case of two equilibrium positions with opposite dipole direction the following four states exist:

i	Configuration	m_i	p_i	energy
1	← →	$m_1 = 0$	$p_1 = (1-w)^2$	0
2	← ←	$m_2 = -2\mu$	$p_2 = w(1-w)$	$V(T)$
3	→ →	$m_3 = 2\mu$	$p_3 = w(1-w)$	$V(T)$
4	→ ←	$m_4 = 0$	$p_4 = w^2$	$2V(T)$

Here the probability p_i was assumed to be the product of the probabilities of finding either of the two dipoles in its respective direction. Thus for $i = 1$ both dipoles are in 'right' positions, each having the probability $(1-w)$. This way of calculating p_i is not free from objections because it does not take into account the correlation between neighbouring dipoles giving rise to short-distance order. It is, however, a sufficiently good approximation for our present qualitative treatment.

With the above four states equation 8.19 becomes

$$\overline{\mathbf{mm}^*} = 8w(1-w)\mu\mu^*, \qquad (8.20)$$

where $+2\mu^*$ is the moment of a spherical region around the unit cell if the moment of the latter is kept at $\pm 2\mu$.

Inserting from 8.20 into 7.39 [it should be remembered that this equation holds for poly-crystalline material], we find by adding $\epsilon_\infty - n^2$ (see above) as contribution of elastic dipole-displacement

$$\epsilon_s - \epsilon_\infty = \frac{3\epsilon_s}{2\epsilon_s + n^2} \frac{4\pi N_0 \mu\mu^*}{3kT} 4w(1-w). \qquad (8.21)$$

Here N_0 is the number of dipoles per unit volume (not the number of unit cells as in 7.39).

For temperatures below the transition point, $w(1-w)$ is probably the determining factor leading to an increase of ϵ_s with temperature. Above the transition point $w = \frac{1}{2}$, i.e. $4w(1-w) = 1$. Equation 8.21 then becomes nearly identical with the Kirkwood equation 8.5 for liquids. A difference remains (i) because 8.21 contains ϵ_∞ instead of n^2 at the left-hand side, and (ii) because μ^* is defined differently.†

A more exact calculation can be carried out, however, above

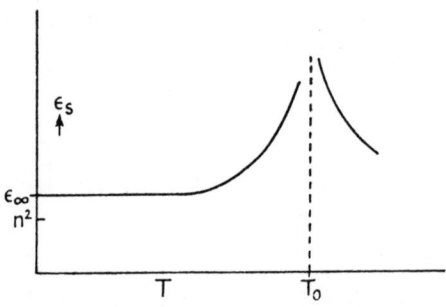

Fig. 10. Temperature-dependence of the dielectric constant of a dipolar solid showing an order-disorder transition, schematically.

T_0 since the unit cell then contains only one dipole. This calculation follows the lines given above for liquids and leads exactly to the Kirkwood formula 8.5. The difference from 8.21 should thus be due to the approximations made in its derivation.

Thus the dielectric constant of a solid (cf. Fig. 10) which has no permanent polarization should start at $T = 0$ with a value ϵ_∞ which is slightly larger than n^2, and should rise very slowly at first and more rapidly as T approaches T_0, provided T_0 is below the melting-point. Above T_0, ϵ_s should decrease with T and there should be no appreciable variation at the melting-point. If, on the other hand, T_0 cannot be reached below the melting-point, then ϵ_s increases up to the melting-point, and then decreases.

The similarity of the dielectric behaviour of disordered solids

† Kirkwood: μ^* is the moment of a spherical region if one of its dipoles is kept in a fixed direction μ/μ. Equation 8.21: $2\mu^*$ is the corresponding moment if two dipoles (of one unit cell) are kept parallel with a total moment 2μ.

and of liquids is not surprising. In fact we have seen before (§§ 4 and 6) that a dipole having two equilibrium positions with equal energy behaves similarly to a dipole with a continuous range of such positions.

Finally, it should be remarked that as for liquids the quantity $|\mu^*-\mu|/\mu$ is a measure for the short-distance order. Above the transition temperature T_0 this quantity can be found experimentally if ϵ_s is known over a sufficiently large temperature range. Near T_0 it may be expected to depend sensitively on the structure.

Intramolecular rotation

Consider briefly the case of a liquid consisting of molecules in which intramolecular rotation is of importance. Again the unit cell contains one single molecule, but its dipole moment $\mathbf{\mu}(x)$ may depend on the set of coordinates x describing intramolecular rotation. Thus we put $\mathbf{m}(x) = \mathbf{\mu}(x)$, and introduce correspondingly $\mathbf{m}^*(x) = \mathbf{\mu}^*(x)$ as moment of spherical region. In the Kirkwood formula 8.5 the term $\mathbf{\mu\mu}^*$ must then be replaced by the average $\overline{\mathbf{\mu}(x)\mathbf{\mu}^*(x)}$. If in particular the molecule has a number of semi-stable states with moments $\mathbf{\mu}_1, \mathbf{\mu}_2,...$, besides the ground state whose moment is $\mathbf{\mu}_0$, then similar to 8.19

$$\overline{\mathbf{\mu}(x)\mathbf{\mu}(x)^*} = \sum_{i \geq 0} \mathbf{\mu}_i \mathbf{\mu}_i^* p_i, \tag{8.22}$$

where p_i represents the probability of finding the molecule in the state i.

CHAPTER III

DYNAMIC PROPERTIES

9. The establishment of equilibrium

THE present chapter will be devoted to an investigation of the dynamic properties of dielectrics. This will be found to be a much more difficult task than the development of the theory of the static properties, and in fact it will be seen that only for dilute solutions of dipolar molecules has it been possible so far to carry out quantitative calculations. The reason for the greater difficulties in dealing with the dynamic properties is evident if we remember that for the static case it was not necessary to investigate the kinetic properties of molecules. In the present section a qualitative discussion of the problems involved will be given.

It will be remembered that in § 2 it was found that in alternating electric fields the electric displacement shows a phase-shift with respect to the electric field. This leads to the introduction of two dielectric constants ϵ_1 and ϵ_2, both of which depend on the frequency $\omega/2\pi$ and which according to 2.8 can be considered as real and imaginary components of a complex dielectric constant $\epsilon(\omega)$. The two quantities ϵ_1 and ϵ_2 are not independent of each other, but can both be derived from a function $\alpha(t)$ of time with the help of equations 2.14 and 2.15. Making use of 2.8, these equations can be written as a single complex equation,

$$\epsilon(\omega) = \epsilon_\infty + \int_0^\infty \alpha(x) e^{i\omega x}\, dx, \qquad (9.1)$$

where x is a variable of integration.

The function $\alpha(t)$, according to § 2, describes the decay of the polarization of a dielectric with time if the external field is suddenly removed. Alternatively the gradual increase of polarization with time to its equilibrium value if the dielectric is brought into a constant field can also be described with the help of the decay function $\alpha(t)$. It is thus seen that with the use of the macroscopic equation 9.1 the frequency dependence of the complex dielectric constant (and hence of dielectric loss, cf. 3.15) is uniquely connected with the way in which equilibrium is established in a dielectric brought into a constant field. We shall choose this latter process for the qualitative considerations which follow because it is very suitable for the purpose of showing in a simple way the difficulties involved.

As in § 4, consider the two characteristic types of displacement of charges, namely (i) elastic displacement, and (ii) displacement to another equilibrium position, and assume that all interaction forces between particles can be neglected. From § 4 we then know the average position of a charge in the presence and in the absence of a field, or rather its average displacement by the field. It is now our task to consider the influence of the field in more detail by assuming that the dielectric consists of such an assembly of (non-interacting) charges which, in the absence of a field, are in thermal equilibrium.

Case (i). *Elastic displacement*

Each particle of mass m and charge e is bound elastically to its equilibrium position and will carry out harmonic oscillations with frequency ω_0 about it. If $\mathbf{r}$ is the displacement, then in the absence of a field

$$\frac{d^2\mathbf{r}}{dt^2} = -\omega_0^2 \mathbf{r}, \qquad (9.2)$$

and hence
$$\mathbf{r} = \mathbf{C}_0 \cos(\omega_0 t + \delta_0), \qquad (9.3)$$

where the maximum amplitude $\mathbf{C}_0$ and the phase δ_0 are independent of time, and the energy is given by $\tfrac{1}{2}m\omega_0^2 C_0^2$. If the motion is not perturbed, the oscillator will keep this same energy indefinitely. It should be realized that this is in contradiction to the postulates of statistical mechanics, according to which within a sufficiently long interval of time the oscillator should be found with various energies according to the Boltzmann theorem, and its average energy at a temperature T should be kT. To achieve this some kind of interaction with other particles or with the surrounding medium—permitting an exchange of energy—must be assumed. Frequently it is assumed that this interaction takes the form of collisions of extremely short duration. This means that the equation of motion 9.2 holds except during a collision. Therefore the solution 9.3 holds between two collisions, but at a collision both the amplitude C_0 and the phase δ_0 usually change their value. The assumption of collisions of very short duration means that each oscillator actually satisfies the equation of motion obtained by neglecting all interaction terms, but as a result of a collision it makes a transition into a state with different energy (and phase). This, as may be shown by statistical mechanics, leads to the correct average energy if averaged over a time which is long compared with the time between two collisions. It is of great importance that in equilibrium this average is independent of the nature of the collisions. Equilibrium properties can, therefore, be derived without consideration of collisions—quite in contrast to the question of attaining equilibrium which we intend to investigate presently. It should be remarked first that the collisions need not occur between the oscillators themselves as one would expect in a gas of such oscillators. They might be considered as dissolved in another medium (liquid or solid) which is in thermal equilibrium, and the collisions to be considered occur between the oscillators and the particles of the medium.

Assume now a constant field **f** to be applied at the time $t = 0$. Then the displacements satisfy the new equation of motion 4.8 instead of 9.2. A particle now oscillates about a new position of equilibrium displaced by a vector $\bar{\mathbf{r}}$ relative to the former one, but the motion is still harmonic (cf. 4.9 and 4.10). Both the maximum amplitude C and the phase δ usually change, however, at $t = 0$, although **r** remains continuous, as is shown in Fig. 11. This assumes, of course, that no collisions occur during the represented time interval.

To obtain the polarization P of a dielectric substance consisting of a large number of oscillators the vector sum of all displacements **r** has to be formed according to 4.6. At a given instant of time $t < 0$ in the absence of a field the phase-angle δ_0 can be assumed to have any value between 0 and 2π with equal probabilities. The vector sum of all displacements, and hence the total electric moment of the dielectric, will, therefore, vanish at any instant of time. Now let us assume as a first approach that no collisions occur after the field has been applied at $t = 0$. Then since all individual displacements vary continuously the same must hold of the polarization P. On the other hand, the

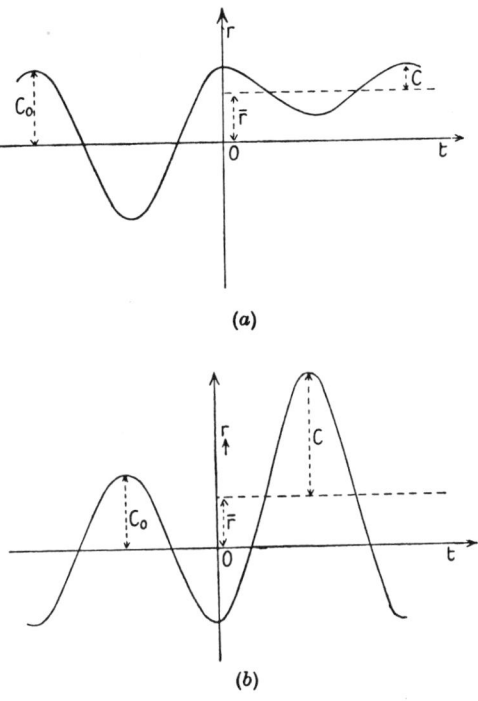

FIG. 11. Time dependence of the displacement r of a harmonic oscillator if a constant field f is applied at the time $t = 0$. The displacement and its time derivative remain continuous at $t = 0$, but the average displacement is altered. Two characteristic cases are shown: (a) both phases before and after application of the field vanish; the maximum amplitude C for $t > 0$ is smaller than it was for $t < 0$; (b) both phases are equal to $-\pi$ and the maximum amplitude is larger for $t > 0$ than for $t < 0$.

time average of the displacement of any single oscillator is equal to $\bar{r}$ so that the average polarization too is different from zero. It will therefore oscillate about its average value with the frequency $\omega_0/2\pi$. This means that after application of the field the oscillators are no longer distributed uniformly over all values of the phase δ, as can also be shown by direct calculation.

Collisions, as has been pointed out above, alter the phases.

They will tend to suppress the oscillations of the polarization about its equilibrium value. One should expect, therefore, that after application of a field the equilibrium polarization is reached steadily if the average time between two collisions of an oscillator

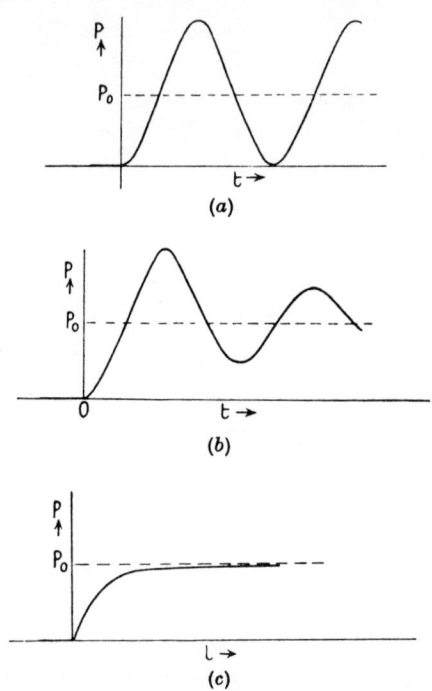

Fig. 12. Time dependence of the polarization of a dielectric if a field is applied at the time $t = 0$. (a) and (b), the material consists of dipolar oscillators, (a) without, (b) with collisions, leading to oscillations of the polarization about its average. (c) The material consists of rigid dipoles with several equilibrium positions.

is shorter than its period. Otherwise transient oscillations of the polarization will be excited which last for a time of the order of the time between two collisions (cf. Fig. 12 a, b, and c).

Case (ii). *Displacement into another equilibrium position*

As in § 4 (cf. Fig. 4), a charged particle is assumed to possess two equilibrium positions A and B at distance b from each other and separated by a potential barrier. In the absence of an

III, § 9   ESTABLISHMENT OF EQUILIBRIUM   67

external field the potential energy of the particle is the same in
A and B. In the presence of a field **f** according to 4.12 there is a
difference in the potential energies at A and B. In an assembly
of a great number of particles of this kind in the absence of a
field an equal number will oscillate about each of the two
equilibrium positions with an average energy kT, if thermal
equilibrium is assumed to exist, and if the height of the potential
barrier is large so that the potential energy H of a particle on top
of it is large compared with kT,

$$H \gg kT. \qquad (9.4)$$

This means that the fraction of particles with sufficient energy
to go over the top of the potential barrier is extremely small.
Its order of magnitude is given by the Boltzmann factor

$$\exp(-H/kT).$$

Assume again that at the time $t = 0$ an electric field **f** is
applied. Then in the absence of collisions the number of particles
oscillating about A and about B is not altered, because the only
action of the field is to alter slightly the equilibrium position—
as pointed out above, case (i). The field is not able, however, to
lift a particle over the potential barrier (cf. § 4). On the other
hand, in equilibrium the number of particles near A is larger by
a fraction of the order ebf/kT (independent of H) than the number
of particles near B, as was shown in § 4. In the absence of
collisions, therefore, equilibrium cannot be established. This
was also found to be so in case (i), but in the present case lack of
equilibrium has a much more serious effect. For in adding all
the displacements in order to calculate the polarization, lack
of collisions meant in case (i) that the polarization was
oscillating about its equilibrium value. In the present case,
however, the required displacements cannot be carried out
without collisions, which means complete lack of polarization of
the required type.

Collisions will tend to establish equilibrium. The time required for this process depends on details of the model. It is often
assumed that collisions on either side of the potential barrier
are very frequent, so that the particles on each side may always

be considered to be in equilibrium amongst themselves. By this it is meant that a particle having sufficient energy to flow over the potential barrier to the other side will suffer so many collisions that it is very unlikely to oscillate back to the side from which it came originally because on an average over many collisions a particle has an energy $kT \ll H$.

Starting now from equilibrium in the absence of a field, the immediate effect of the application of a field in the A–B direction is to lift the potential near A by the amount $e\mathbf{fb}$. As a consequence the fraction of particles with sufficient energy to move over the potential barrier is approximately given by

$$e^{-(H-e\mathbf{bf})/kT} \text{ and } e^{-H/kT}$$

for the $A \to B$ and the $B \to A$ directions respectively,† because the height of the potential barrier measured from A is now only $H - e\mathbf{bf}$. Thus if $\omega_0/2\pi$ is the frequency of oscillation of a particle, the probability per second for the transfer of a particle from B to A is given by

$$w_{21} = \frac{\omega_0}{2\pi} e^{-H/kT}, \qquad (9.5)$$

and for the transition $A \to B$ considering 4.18 it is

$$w_{12} = \frac{\omega_0}{2\pi} e^{-(H-e\mathbf{fb})/kT} \simeq \frac{\omega_0}{2\pi} e^{-H/kT}\left(1 + \frac{e\mathbf{fb}}{kT}\right) = w_{21}\left(1 + \frac{e\mathbf{fb}}{kT}\right). \qquad (9.6)$$

Thus if at any instant of time there is a number of particles $N_1(t)$ at A and a number $N_2(t)$ at B, a number $N_1 w_{12}$ will flow per second from A to B and a number $N_2 w_{21}$ from B to A. Therefore the rates of change of N_1 and N_2 respectively are given by

$$\frac{dN_1}{dt} = -N_1 w_{12} + N_2 w_{21} \qquad (9.7)$$

$$\frac{dN_2}{dt} = -N_2 w_{21} + N_1 w_{12} = -\frac{dN_1}{dt}. \qquad (9.8)$$

† By introducing a different normalization of the potential of the external field the two exponents can be written in a more symmetrical way as $H - \tfrac{1}{2}e(\mathbf{bf})$ and $H + \tfrac{1}{2}e(\mathbf{bf})$, leaving the results unaltered.

It follows that the total number N of particles,
$$N = N_1 + N_2, \tag{9.9}$$
is independent of time, as required, because
$$\frac{dN_1}{dt} + \frac{dN_2}{dt} = \frac{dN}{dt} = 0. \tag{9.10}$$
Subtracting 9.7 from 9.8 with the use of 9.9 yields
$$\frac{d}{dt}(N_2 - N_1) = -(w_{12} + w_{21})(N_2 - N_1) + (w_{12} - w_{21})N. \tag{9.11}$$

Now according to 9.6 and 4.18,
$$w_{12} + w_{21} = 2w_{21}\left(1 + \frac{1}{2}\frac{e\mathbf{fb}}{kT}\right) \simeq 2w_{21}, \tag{9.12}$$
$$w_{12} - w_{21} = \frac{e\mathbf{fb}}{kT}w_{21}. \tag{9.13}$$

Using these two equations, 9.11 becomes
$$\frac{1}{2w_{21}}\frac{d}{dt}(N_2 - N_1) = -(N_2 - N_1) + \frac{1}{2}\frac{e\mathbf{fb}}{kT}N. \tag{9.14}$$

If we assume
$$N_1(0) = N_2(0) = \tfrac{1}{2}N$$
at $t = 0$, 9.14 is solved by
$$N_2 - N_1 = \frac{N}{2}\frac{e\mathbf{fb}}{kT}(1 - e^{-2w_{21}t}). \tag{9.15}$$

The induced polarization is proportional to $N_2 - N_1$ and thus approaches its equilibrium value exponentially, as shown in Fig. 12c.

In contrast to case (i), equilibrium is reached in the present case by the influence of the field on the probability of transfer of a particle from A to B and B to A respectively, according to equations 9.5 and 9.6 or 9.12 and 9.13. This model can be generalized by introducing transition probabilities w_{12} and w_{21} measuring the probabilities per second for the transition of a particle from A to B and from B to A respectively, without specifying them by equations 9.5 and 9.6. This means that equation 9.11 holds as before. Instead of calculating w_{12} and

w_{21} from a model, and then deriving 9.12 and 9.13, these equations can also be obtained from 9.11 by considering solutions corresponding to equilibrium which can be obtained from § 4 without the knowledge of the transition probabilities. Thus since in equilibrium $d(N_1-N_2)/dt = 0$, in the absence of a field $N_1 = N_2$, and hence from 9.11 $w_{12} = w_{21}$. In the presence of a field, using 4.13, 4.12, and 4.18,

$$N_2 - N_1 = N(p_B - p_A) = \frac{N}{2} \frac{e\mathbf{fb}}{kT}. \tag{9.16}$$

This must be a solution of 9.11 if $N_2 - N_1$ is independent of time, which leads to 9.12 and 9.13.

The model discussed above can be further generalized by considering dipoles which can have several equilibrium positions with certain transition probabilities between them. In equilibrium in the absence of a field the number of transitions from a given position is just balanced by the number of transitions from other positions to it. An electric field alters the transition probabilities, and hence, in equilibrium, the distribution of dipoles over the various positions. The time required to attain equilibrium depends on the transition probabilities.

The preceding considerations have thus shown that in the case of elastic binding the field displaces the charges which then oscillate about their equilibrium positions. In the case in which charged particles or dipoles possess several equilibrium positions the field does not, by immediate action on the charges, transfer them to their new positions, but it alters the transition probabilities between them. This in turn leads to the establishment of equilibrium.

10. The Debye equations

In the present section it is intended to derive equations for the frequency dependence of the complex dielectric constant $\epsilon(\omega)$ which should hold for the case of dilute solutions of dipoles in liquids and solids and for a few other cases. These equations were first established by Debye [D2] and subsequently have been applied to many substances, not always, unfortunately, with the necessary discrimination with respect to the intended range of validity. We shall base the considerations of this section on the hypothesis that in constant external fields equilibrium is attained exponentially with time, as corresponds to case (ii) of § 9. We thus assume for the decay function $\alpha(t)$,

$$\alpha(t) \propto e^{-t/\tau}, \tag{10.1}$$

where τ is independent of time but may depend on temperature. That this assumption leads to the required properties can easily be shown from the relationship 2.12 between the electric field $E(t)$ and the electric displacement $D(t)$, both of which may depend on time. In 2.12 it was assumed that both D and E vanish for times $t < 0$. If this is no longer the case, then 2.12 has to be replaced by

$$D(t) = \epsilon_\infty E(t) + \int_{-\infty}^{t} E(u)\alpha(t-u)\,du. \tag{10.2}$$

This integral equation can easily be transformed into a differential equation. For on differentiating 10.2 with respect to the time and making use of

$$\frac{d\alpha(t)}{dt} = -\frac{1}{\tau}\alpha(t), \tag{10.3}$$

which follows from 10.1, we find after multiplication by τ

$$\tau\frac{dD(t)}{dt} = \epsilon_\infty \tau\frac{dE(t)}{dt} + \tau\alpha(0)E(t) - \int_{-\infty}^{t} E(u)\alpha(t-u)\,du. \tag{10.4}$$

Adding 10.2 and 10.4 yields

$$\tau\frac{d}{dt}(D - \epsilon_\infty E) + (D - \epsilon_\infty E) = \tau\alpha(0)E. \tag{10.5}$$

To determine the constant $\alpha(0)$ consider the special case of equilibrium in a constant field. This means that

$$\frac{d}{dt}(D - \epsilon_\infty E) = 0, \qquad D = \epsilon_s E,$$

and hence from 10.5

$$\tau\alpha(0) = \epsilon_s - \epsilon_\infty. \tag{10.6}$$

Therefore inserting 10.6 into 10.5,

$$\tau \frac{d}{dt}(D - \epsilon_\infty E) + (D - \epsilon_\infty E) = (\epsilon_s - \epsilon_\infty) E \qquad (10.7)$$

is found as the differential equation connecting $D(t)$ with $E(t)$ on the assumption that the decay function $\alpha(t)$ is given by (cf. 10.1 and 10.6)

$$\alpha(t) = \frac{\epsilon_s - \epsilon_\infty}{\tau} e^{-t/\tau}. \qquad (10.8)$$

We shall now use equation 10.7 in the investigation of the approach to equilibrium of a condenser. The following two cases are to be considered:

(a) Constant charge on the condenser plates. Then

$$\frac{dD}{dt} = 0, \quad D = D_0,$$

and hence, using 10.7,

$$\tau' \frac{dE}{dt} + E = \frac{D_0}{\epsilon_s}, \quad \text{i.e.} \quad D_0 - \epsilon_s E \propto e^{-t/\tau'}, \qquad (10.9)$$

where

$$\tau' = \frac{\epsilon_\infty}{\epsilon_s} \tau. \qquad (10.10)$$

(b) Constant voltage at the condenser plates, i.e.

$$\frac{dE}{dt} = 0, \quad E = E_0.$$

It follows with 10.7 that

$$\tau \frac{dD}{dt} + D = \epsilon_s E_0, \quad \text{i.e.} \quad D - \epsilon_s E_0 \propto e^{-t/\tau}. \qquad (10.11)$$

Both cases thus lead to exponential approach to equilibrium.

In periodic fields assume E to be represented by equation 2.9, i.e. $E \propto \exp(-i\omega t)$. Then, introducing the complex dielectric constant ϵ (cf. 2.8), we find, using 2.10,

$$\frac{dE}{dt} = -i\omega E, \quad D = \epsilon(\omega) E, \quad \frac{dD}{dt} = -i\omega \epsilon(\omega) E. \qquad (10.12)$$

Introducing this into 10.7 leads to

$$\epsilon(\omega) - \epsilon_\infty = \frac{\epsilon_s - \epsilon_\infty}{1 - i\omega \tau}. \qquad (10.13)$$

An alternative way of deriving this equation is to insert 10.8 into 9.1,

$$\epsilon(\omega)-\epsilon_\infty = (\epsilon_s-\epsilon_\infty)\frac{1}{\tau}\int_0^\infty e^{i\omega x - x/\tau}\,dx, \qquad (10.14)$$

which after integration results in equation 10.13.

Separating the real and imaginary parts in 10.13 according to 2.8 we find

$$\epsilon_1(\omega)-\epsilon_\infty = \frac{\epsilon_s-\epsilon_\infty}{1+\omega^2\tau^2}, \qquad (10.15)$$

$$\epsilon_2(\omega) = \frac{(\epsilon_s-\epsilon_\infty)\omega\tau}{1+\omega^2\tau^2}, \qquad (10.16)$$

and for the loss-angle ϕ, using 2.5,

$$\tan\phi = \frac{\epsilon_2}{\epsilon_1} = \frac{(\epsilon_s-\epsilon_\infty)\omega\tau}{\epsilon_s+\epsilon_\infty\omega^2\tau^2}. \qquad (10.17)$$

Equations 10.15–10.17 (also 10.13) will be denoted as the Debye equations and the constant τ will be called the relaxation time. They describe the properties of a dielectric substance in alternating fields on the assumption of an exponential decay function $\alpha(t)$ (cf. 10.8). Some models leading to such a decay function will be studied in § 11. Most of them require

$$\epsilon_s-\epsilon_\infty \ll 1, \qquad (10.18)$$

a condition which normally is fulfilled in dilute solutions only.

In discussing the properties of the Debye equations it should be noted that the dielectric constants ϵ_1 and ϵ_2 depend on at least two parameters, the angular frequency ω and the temperature T. The frequency dependence is expressed explicitly, but the temperature appears implicitly through $\epsilon_s-\epsilon_\infty$ and τ, both of which usually depend on T. They may depend on other parameters as well, variations of which will not be considered here. For the following it will be assumed that both ϵ_s and ϵ_∞ are known as functions of T. If τ were known as well, a new variable

$$z = \log\omega\tau = \log\omega + \log\tau \qquad (10.19)$$

could be introduced in terms of which 10.15 and 10.16 become

$$\frac{\epsilon_1-\epsilon_\infty}{\epsilon_s-\epsilon_\infty} = \frac{1}{1+e^{2z}} = \frac{e^{-z}}{e^z+e^{-z}}, \qquad \frac{\epsilon_2}{\epsilon_s-\epsilon_\infty} = \frac{1}{e^z+e^{-z}}. \qquad (10.20)$$

Fig. 13 shows these functions. It will be noted that $\epsilon_2/(\epsilon_s-\epsilon_\infty)$ is a symmetrical function of z.

Actually τ cannot be assumed to be known but must be determined from measurements of ϵ_1 and ϵ_2 at various frequencies and temperatures. Assuming that the Debye equations are fulfilled $\tau(T)$ can, however, easily be found from the frequency

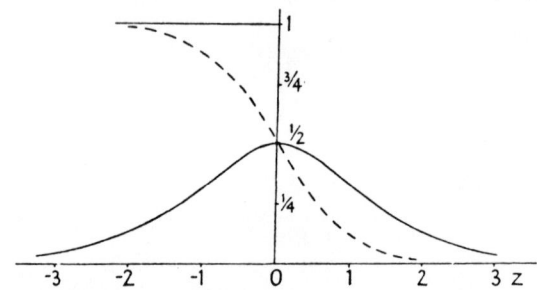

Fig. 13. The Debye functions $\epsilon_2/(\epsilon_s-\epsilon_\infty)$ (full line), and $(\epsilon_1-\epsilon_\infty)/(\epsilon_s-\epsilon_\infty)$ (dotted line), according to equations 10.20.

at which ϵ_2 has its maximum. In fact at constant temperature the angular frequency ω_m of this maximum is determined by

$$\frac{\partial \epsilon_2}{\partial \omega} = 0, \quad \text{if} \quad \omega = \omega_m \text{ and } T = \text{constant}. \qquad (10.21)$$

Hence using 10.16,

$$\omega_m(T) = \frac{1}{\tau(T)}. \qquad (10.22)$$

Inserting this value into 10.15–10.17 we find for the dielectric constants and the loss angle

$$\epsilon_1 = \tfrac{1}{2}(\epsilon_s+\epsilon_\infty), \qquad \epsilon_2 = \tfrac{1}{2}(\epsilon_s-\epsilon_\infty), \qquad \tan\phi = \frac{\epsilon_s-\epsilon_\infty}{\epsilon_s+\epsilon_\infty},$$

$$\text{if } \omega = \omega_m. \qquad (10.23)$$

Thus from the frequency at which ϵ_2 has its maximum we find τ, and from its height we ought to obtain $\epsilon_s-\epsilon_\infty$ if the Debye equations are fulfilled.

Instead of considering the maximum of ϵ_2 it is often preferred

THE DEBYE EQUATIONS

to determine the angular frequency ω_ϕ at which the loss-angle ϕ has its maximum. We then require

$$\frac{\partial \tan\phi}{\partial \omega} = 0, \text{ if } \omega = \omega_\phi \text{ and } T = \text{constant}. \quad (10.24)$$

Using 10.17, this yields

$$\omega_\phi = \frac{1}{\tau}\sqrt{\frac{\epsilon_s}{\epsilon_\infty}}. \quad (10.25)$$

In view of 10.18, ω_ϕ and ω_m will be nearly equal for most substances for which the Debye equations can be expected to hold. Inserting 10.25 into 10.15–10.17 we find

$$\epsilon_1 = 2\frac{\epsilon_s \epsilon_\infty}{\epsilon_s + \epsilon_\infty}, \qquad \epsilon_2 = \frac{\epsilon_s - \epsilon_\infty}{\epsilon_s + \epsilon_\infty}\sqrt{(\epsilon_s \epsilon_\infty)},$$

$$\tan\phi = \frac{\epsilon_s - \epsilon_\infty}{2\sqrt{(\epsilon_s \epsilon_\infty)}}, \text{ if } \omega = \omega_\phi. \quad (10.26)$$

It is an interesting feature of equations 10.23 and 10.26 that the values of ϵ_1 and ϵ_2 at the frequency ω_m or ω_ϕ at which either ϵ_2 or $\tan\phi$ has its maximum are independent of this frequency and of the relaxation time, and are expressible by the static dielectric constant ϵ_s and the high-frequency dielectric constant ϵ_∞.

We shall use this occasion to give a more precise definition of ϵ_∞ than that given previously. According to the Debye equations ϵ_1 decreases from ϵ_s to ϵ_∞ in the frequency region in which ϵ_2 has relatively large values. Therefore ϵ_∞ is the value which is asymptotically approached by $\epsilon_1(\omega)$ at frequencies sufficiently larger than ω_m to make $\epsilon_2(\omega)$ relatively small. This value of ϵ_∞ need not coincide with the optical dielectric constant because most substances absorb in the infra-red region (see also § 13).

If $\epsilon_s - \epsilon_\infty$ is very small, as is required by 10.18, then ϵ_s, ϵ_∞, and ϵ_1 are nearly equal,

$$\epsilon_s \simeq \epsilon_\infty \simeq \epsilon_1(\omega), \quad (10.27)$$

and very precise measurements would be required to obtain

$\epsilon_s - \epsilon_\infty$. This quantity can, however, be obtained with the help of the phenomenological relation 2.18 if

$$\epsilon_2(\omega) = \epsilon_1 \tan\phi \simeq \epsilon_s \tan\phi$$

is known for all frequencies for which it has appreciable values. For on inserting ϵ_2 from 10.16, equation 2.18 becomes

$$\epsilon_s - \epsilon_\infty = \frac{2}{\pi}\int_0^\infty \epsilon_2(\omega)\frac{d\omega}{\omega} = (\epsilon_s - \epsilon_\infty)\frac{2}{\pi}\int_0^\infty \frac{\omega\tau}{1+\omega^2\tau^2}\frac{d\omega}{\omega}, \tag{10.28}$$

which is an identity because

$$\int_0^\infty \frac{\omega\tau}{1+\omega^2\tau^2}\frac{d\omega}{\omega} = \frac{\pi}{2}.$$

In the particular case of dilute solutions of dipolar molecules in non-polar liquids $\epsilon_s - \epsilon_\infty$ can be inserted from 6.24. In this case the dielectric constant ϵ_0 of the solvent is nearly equal to ϵ_∞, i.e. with 10.27,

$$\epsilon_0 \simeq \epsilon_\infty \simeq \epsilon_1(\omega) \simeq \epsilon_s. \tag{10.29}$$

Thus from 6.24, 10.16, 10.17, and 10.29

$$\epsilon_s \tan\phi \simeq \epsilon_2 \simeq \frac{4\pi\mu_v^2 N_0}{3kT}\left(\frac{\epsilon_s+2}{3}\right)^2\left\{1 - \frac{2(\epsilon_s-1)(\epsilon_s-n^2)}{(2\epsilon_s+n^2)(\epsilon_s+2)}\right\}^2 \frac{\omega\tau}{1+\omega^2\tau^2}, \tag{10.30}$$

for spherical molecules and $\epsilon_s - \epsilon_\infty \ll 1$.

Here n is the refractive index of a pure liquid of the dissolved molecules. The factor containing n is not of great importance because $\{\quad\}^2$ usually differs from unity by a few per cent. only.

In the above discussion of the Debye equations we have been mainly concerned with the frequency dependence of the dielectric constants using the temperature as a parameter. In practice ϵ_1 and ϵ_2 are often measured as functions of the temperature T with the frequency as a parameter. This does not permit such an immediate comparison with the Debye equations because the latter do not depend on T explicitly. Assuming that $\epsilon_s - \epsilon_\infty$ is known as a function of T, it is advisable in this case to plot the function $\epsilon_2/(\epsilon_s-\epsilon_\infty)$ against T. Its maximum is determined by (cf. 10.16)

$$0 = \frac{\partial}{\partial T}\left(\frac{\epsilon_2}{\epsilon_s-\epsilon_\infty}\right) = \frac{\partial}{\partial T}\frac{\omega\tau}{1+\omega^2\tau^2} = \frac{\partial}{\partial\tau}\left(\frac{\omega\tau}{1+\omega^2\tau^2}\right)\frac{d\tau}{dT}, \quad \text{i.e.} \quad \omega\tau = 1$$

as before 10.22. This means that the maximum for a given frequency occurs at a temperature T_m at which

$$\tau(T_m) = 1/\omega. \tag{10.31}$$

In the case that ϵ_1 and ϵ_2 are measured as functions of ω for a given temperature T, the Debye equations can be checked immediately once $\tau(T)$ has been found from the position of the maximum of ϵ_2 according to 10.22. For τ does not depend on ω, and its value can thus be inserted into equations 10.15–10.17 which then contain no unknown quantities. If, on the other hand, ϵ_1 and ϵ_2 are measured for a given frequency as functions of T, the value found for $\tau(T_m)$ must not be introduced into 10.15–10.17 because τ varies with T. Now if 10.18 is fulfilled and thus an exact measurement of $\epsilon_s-\epsilon_\infty$ is difficult, knowledge of the temperature dependence of ϵ_2 at a given frequency does not permit a check of the Debye equations. If, however, $\epsilon_s-\epsilon_\infty$ and $\epsilon_1-\epsilon_\infty$ are known as functions of the temperature, $\omega\tau$ can be derived from 10.15, i.e.

$$(\omega\tau)^2 = \frac{\epsilon_s-\epsilon_1}{\epsilon_1-\epsilon_\infty}, \tag{10.32}$$

and then inserted into 10.16, resulting in

$$\epsilon_2 = \surd\{(\epsilon_s-\epsilon_1)(\epsilon_1-\epsilon_\infty)\}. \tag{10.33}$$

Equations 10.32 and 10.33 are equivalent to the Debye equations 10.15 and 10.16 in their original form. They have the advantage of showing clearly that the relaxation time τ is a quantity which can be calculated from the measurable quantities according to 10.32, whereas 10.33 is a relation between measurable quantities only. Their disadvantage is that they require measurements of $\epsilon_s-\epsilon_1$ and $\epsilon_1-\epsilon_\infty$ which in the cases where the Debye equations are supposed to hold are very small (cf. 10.18 and 10.27). Whenever this is possible, however, 10.33 affords a very simple test of the Debye equations whether temperature, frequency, or both are varied.

11. Models for the Debye equations

In the previous section it was shown that an exponential decay function leads to the Debye equations. In the present section, models for which these equations hold will be studied and expressions for the relaxation time will be obtained. A very simple model has already been discussed in case (ii) of § 9 and § 4. In this model the dielectric material contains an assembly of charged particles whose interaction can be neglected. Each charge has two equilibrium positions separated by a high potential barrier. It is then assumed that each particle collides with the surrounding medium and that the number of collisions per second is such that the average time τ_0 between two collisions is small compared with the average time τ which a particle spends near one of its equilibrium positions before jumping to the other one. This leads to the linear differential equation 9.11 for the difference of the number of particles occupying the two positions. Its solution 9.15 approaches its equilibrium value exponentially with a relaxation time (cf. 9.5)

$$\tau = \frac{1}{2w_{21}} = \frac{\pi}{\omega_0} e^{H/kT}, \qquad H \gg kT, \qquad \tau_0 \ll \tau, \qquad (11.1)$$

where H is the height of the potential barrier and $\omega_0/2\pi$ the frequency of oscillation about either of the equilibrium positions. For this model, according to 9.5 and 9.6, the transition probabilities w_{12} and w_{21} of a particle between the two equilibrium positions are equal in the absence of a field, but are altered slightly when an external field is applied.

It should be pointed out that from the assumptions made it follows that the derived exponential law can be true only on an average over time intervals which are large compared with τ_0. Therefore the derivation of the Debye equations from this law can hold only if the period of the field is large compared with τ_0. It must, therefore, be concluded that this model leads to the Debye equations only when the following conditions hold:

$$\tau_0 \ll \tau, \qquad \tau_0 \ll 1/\omega. \qquad (11.2)$$

Thus, in order to ascertain whether or not the Debye equations hold in the main region of absorption (near $\omega\tau \sim 1$), it would be necessary to determine τ_0, and this in turn would require a detailed knowledge of the interaction between the charges and their surroundings. It will also be seen from conditions 11.2 that even if the Debye equations do hold in this main region deviations may be expected at higher frequencies.

It is important to realize that an exponential decay function (and hence the Debye equations) can only be obtained when the particles can be considered as independent of each other. Interaction between particles (which is to be distinguished from that between a particle and its surroundings) would mean that the transition probabilities are not constant, but depend on the positions of the neighbours. Hence equations 9.7 and 9.8 would be no longer linear and, therefore, could not be solved by exponential functions.

Dipolar solids

Models more directly applicable to solids, but leading to a mathematical treatment similar to the above, can easily be devised. Consider, for instance, the model for dipolar solids used in § 8. It consists of dipolar molecules, each of which, owing to the crystalline field, has a number of equilibrium positions with different dipole directions, which are separated by potential barriers. In the simplest case only two equilibrium positions with opposite dipole directions exist. In such a model at low temperatures the dipoles, because of interaction with each other, form an ordered arrangement. At a temperature T_0 an order-disorder transition occurs and for $T > T_0$ long-distance order vanishes. Short-distance order, i.e. order relative to neighbours persists, however, and can be neglected only at still higher temperatures. It will be shown that at these temperatures the Debye equations hold. They should be invalid at temperatures near T_0, but become valid again if $T \ll T_0$.

Let us consider first the high-temperature region. Then in the absence of a field the lowest energy-level of a dipole is the same for both equilibrium directions. To carry out a transition between them a minimum energy, say H, is required to lift the molecule over the potential barrier separating the two equilibrium positions. In the calculation of the probability of a transition between the two positions equation 9.5 cannot now be applied because in general a molecule is also capable of internal excitation, a fact which was not considered in § 9. We shall take account of this by introducing a factor A which only varies slowly with temperature. Then if $1/2\tau$ is the transition probability, 11.1 will be replaced by

$$\tau = \frac{\pi}{2\omega_a} A e^{H/kT}, \tag{11.3}$$

where π/ω_a is the average time required by an excited molecule to turn from one equilibrium direction to the other. In the presence of a field $\mathbf{f}$ the molecule has an additional energy $+\mathbf{\mu f}$ and $-\mathbf{\mu f}$ for the two equilibrium directions of its dipole respectively. The minimum energy required by the molecule to enable it to carry out a transition in the presence of a field is, therefore, $H \pm \mathbf{\mu f}$. Thus, assuming

$$\mu f/kT \ll 1, \tag{11.4}$$

the two transition probabilities are given by

$$w_{12} = \frac{1}{2\tau} e^{\mu f/kT} \simeq \frac{1}{2\tau}\left(1 + \frac{\mathbf{\mu f}}{kT}\right), \tag{11.5}$$

$$w_{21} = \frac{1}{2\tau} e^{-\mu f/kT} \simeq \frac{1}{2\tau}\left(1 - \frac{\mathbf{\mu f}}{kT}\right). \tag{11.6}$$

respectively. Hence if N_1 and N_2 are the numbers of dipoles in the two directions, equations 9.7, 9.8, and 9.11 still hold, provided these new values of w_{12} and w_{21} are used. Therefore inserting 11.5 and 11.6 into 9.11 yields

$$\frac{d}{dt}(N_2 - N_1) = -\frac{1}{\tau}(N_2 - N_1) + \frac{1}{\tau}\frac{\mathbf{\mu f}}{kT} N. \tag{11.7}$$

For constant fields this equation leads to exponential approach to equilibrium and hence according to § 10 to the Debye equations with a relaxation time τ.

Assuming that a molecule interchanges energy very quickly with its surroundings, it can be considered as passing through a number of levels above the energy H during the transition between the two equilibrium positions. A quasi-thermal equilibrium of the excited molecule with its surroundings is therefore established. Under these conditions the probability of finding a molecule with an energy above H is approximately $(D_H/D_0)\exp(-H/kT)$, where D_0 is the number of energy-levels in a range of the order kT near the ground level and D_H is the corresponding number in the excited state above H. Now the transition probability $1/2\tau$ must be equal to this quantity multiplied by π/ω_a, which is the average time required by an excited molecule to turn by an angle π, and divided by 2 because only half of the molecules move in the required direction. Hence by comparison with equation 11.3 it is found that approximately

$$A = \frac{D_0}{D_H}. \tag{11.8}$$

Expression 11.3 for the relaxation time is sometimes (e.g. Frank, F2; Kauzmann, K1) related to similar expressions for the rate of unimolecular chemical reactions. In general a formula of this type will be obtained for any process which requires excitation to an energy H, but calculation of the absolute value of a rate of reaction (requiring the factor A/ω_a) is difficult. The first calculation of this kind is due to Pelzer and Wigner [P4]. Usually ω_a can be assumed to be of the order

$$\omega_a \sim 10^{12}\text{--}10^{14} \text{ per second}, \tag{11.9}$$

so that measurement of the temperature-dependence of τ leads to semi-empirical values for the order of magnitude of A (11.8). Sometimes (e.g. Eyring, E1, E2) it is assumed that the translational or rotational motion of the excited molecule can be separated from the other types of motion. In this case if $\hbar\omega_a \ll kT$, there are about $kT/\hbar\omega_a$ energy-levels, in an interval kT, connected with the rotation of the molecule as a whole. Denoting by D_H^+ the number of energy-levels due to internal excitation of the molecule, we then obtain, using 11.8,

$$D_H \simeq D_H^+ \frac{kT}{\hbar\omega_a}, \quad \text{i.e.} \quad \frac{A}{\omega_a} \simeq \frac{D_0}{D_H^+} \frac{\hbar}{kT}. \tag{11.10}$$

In view of the assumptions that have been made in the derivation of this expression one should not expect it to give more than an order of magnitude. And if $\hbar\omega_a > kT$, a result might be obtained which is not even of the correct order (cf. Pelzer, P3).

Let us now consider our model at lower temperatures. In the neighbourhood of the transition temperature T_0 interaction between dipoles becomes important. The energy of a molecule then depends on the dipole directions of its neighbours which invalidates the assumptions which led to the Debye equations. It will be shown presently, however, that this is no longer the case when the temperature is well below the transition point and the dipoles are therefore in an ordered state. As pointed out in § 8, the lattice then contains two types of sites; and if the state is completely ordered, these sites are occupied by molecules with opposite dipole directions. They may form, for instance, the centres and the corners of a body-centred cubic lattice. Each molecule has a second equilibrium position with opposite dipole direction in which its average energy is higher by an amount $V(T)$ than in the original position. If the temperature is sufficiently low (well below T_0), it is possible to replace $V(T)$ by $V(0)$, the value of V when $T = 0$. It is in this region that we can assume the energy of a dipole to depend on its own direction only, and not on the directions of its neighbours, because the latter are nearly always in the state of lowest energy (ground state). Hence we expect the Debye equations to hold.

To prove this it is sufficient to show that equilibrium is approached exponentially with time since, as shown in § 10, the Debye equations then obtain. It will simplify the calculations if we assume at a given time the existence of a dipole moment of the substance in the absence of a field and calculate its variation with time. Let us distinguish quantities referring to the two types of site by the use of $+$ and $-$ indices. Thus, for instance, w_{12}^+ is the probability per second for a transition of a dipole on a $+$ site from the '1' into the '2' direction. Similarly N_1^+ and N_1^- are the numbers of dipoles in the '1'-direction for the two types of site. The total dipole moment is proportional to the quantity

$$\Delta N = N_1^+ - N_2^+ + N_1^- - N_2^-. \qquad (11.11)$$

Also if N is the total number of dipoles
$$N_1^+ + N_2^+ = N_1^- + N_2^- = \tfrac{1}{2}N. \tag{11.12}$$
Furthermore, in the absence of a field the '1' ('2')-direction plays the same role for $+$sites as the '2' ('1')-direction for $-$sites. Thus in equilibrium at $T = 0$, $N_1^+ = N_2^- = \tfrac{1}{2}N$, and $N_1^- = N_2^+ = 0$. Also
$$w_{12}^+ = w_{21}^-, \qquad w_{21}^+ = w_{12}^-. \tag{11.13}$$
Thus by the same method as that used in deriving equations 9.7 and 9.8 we now find
$$\frac{dN_1^+}{dt} = -\frac{dN_2^+}{dt} = -w_{12}^+ N_1^+ + w_{21}^+ N_2^+, \tag{11.14}$$
from which, with the help of 11.12, we obtain
$$\frac{d}{dt}(N_1^+ - N_2^+) = -2w_{12}^+ N_1^+ + 2w_{21}^+ N_2^+$$
$$= -(w_{12}^+ + w_{21}^+)(N_1^+ - N_2^+) + \tfrac{1}{2}(w_{21}^+ - w_{12}^+)N. \tag{11.15}$$
The same equation holds for the $-$ sites, if we exchange $+$ for $-$ indices. Using equation 11.13 this means that
$$\frac{d}{dt}(N_1^- - N_2^-) = -(w_{12}^+ + w_{21}^+)(N_1^- - N_2^-) - \tfrac{1}{2}(w_{21}^+ - w_{12}^+)N. \tag{11.16}$$
Therefore by adding 11.15 and 11.16 and making use of 11.11 we find
$$\frac{d}{dt}\Delta N = -(w_{12}^+ + w_{21}^+)\Delta N, \quad \text{i.e.} \quad \Delta N \propto e^{-(w_{12}^+ + w_{21}^+)t}, \tag{11.17}$$
which proves that ΔN approaches its equilibrium value ($\Delta N = 0$) exponentially.

Dipolar liquids

It was shown in § 8 that when a disordered solid melts there should be no appreciable change in the static dielectric constant because in both the liquid and solid phases the average energy of a dipole is the same for all equilibrium directions. The fact that—in contrast to liquids—there is only a discrete number of equilibrium directions in solids has no influence on the static dielectric constant.

There is, however, an essential difference in the dynamic properties of solids and liquids. In solids in view of the interaction of a molecule with its neighbours a dipole has a number of equilibrium directions. They are separated by potential barriers over which the dipole must pass in turning from one such direction to another. In liquids the average distance between neighbours, and hence the interaction, is about the same as in solids. Therefore, if the positions of all molecules but one could suddenly be fixed, this selected molecule would behave similarly to a dipolar molecule in a solid, i.e. it would probably possess a number of equilibrium directions separated by potential barriers. It is an essential property of a liquid, however, that its molecules have no fixed positions. Thus, if we imagine one molecule to be turned out of its momentary equilibrium directions, its neighbours will tend to rearrange themselves in such a way as to make this new direction an equilibrium position. So we may have two conceptions of the way in which a dipole in a liquid alters its direction. It may jump into a different direction in a similar way as in solids. This requires—at least for the duration of this jump—that the arrangement of its neighbours should remain unaltered. The second possibility is that such jumps may occur very rarely and that a dipole may alter its direction only in conjunction with a rearrangement of the positions of its neighbours. In fact a dipole might be considered as fixed fairly rigidly relative to its neighbours. The rotation of a dipole would then affect the motion of molecules at some distance from it.

The average motion of these neighbouring molecules might be described by replacing them by a continuous medium with the properties of a macroscopic viscous fluid. This possibility leads to the model used by Debye [*D2*] in which a dipolar molecule is considered to be a sphere of radius a moving in a continuous viscous fluid with viscosity η and obeying the macroscopic equations of flow. The fluid is considered to adhere to the surface of the molecule. On these assumptions the frictional constant ξ of the sphere is given by Stokes's law,

$$\xi = 8\pi\eta a^3. \tag{11.18}$$

This means that when an electric field **f**, making an angle θ with the direction of the dipole **μ**, is applied,

$$\xi \frac{d\theta}{dt} = -\mu f \sin\theta, \qquad (11.19)$$

if no other forces act on the molecule. In this equation it is also assumed that inertia effects of the dipole—which macroscopically would lead to an additional term $I d^2\theta/dt^2$ ($I =$ moment of inertia) on the left-hand side—can be omitted. This omission is justified in so far as this form of the term is concerned since it would hardly be expected that such effects could be described adequately by this macroscopic expression.

The macroscopic equation of flow 11.19 can yield correct values for θ only if account is taken of thermal fluctuations due to the interchange of energy between the molecule and its surroundings. For the above model the energy of a molecule is the kinetic energy of rotation of the sphere and has an average value of order kT. This rotation—corresponding to Brownian motion in translation—changes its direction and magnitude very frequently due to collisions with the molecules of the surrounding liquid. However, if no external forces act on the dipole its average displacement from a given direction vanishes because displacements in all directions have equal probability. The mean square of these displacements, on the other hand, will increase steadily with time.

Equation 11.19 may, therefore, be considered to hold for the average value of θ (for a single dipole) when a field is applied. The actual value of θ may, however, only be expected to be near its average value as long as the mean square displacement due to thermal fluctuations remains a small quantity. For a macroscopic dipole this should practically always be so. According to 11.19 such a dipole will, therefore, gradually approach the direction $\theta = 0$ parallel to the field. In equilibrium it will remain in this direction apart from small fluctuations.

It is to be noted that the value θ used in 11.19, and assumed above to be the time average of the θ-values of a single dipole, may also be considered as the average of such angular displacements of a large number of (non-interacting) dipoles all having the same θ-value within a small range.

In contrast to a macroscopic dipole the fluctuations of a molecular dipole are very large. This follows at once from consideration of the equilibrium distribution of an assembly of dipoles which, according to 6.9 (replacing E by f), is proportional to $\exp(\mu f \cos\theta/kT)$. Since $\mu f \ll kT$ for all practicable fields (cf. § 4), the field has only a slight influence on the distribution function. In other words, a dipole has only a slightly higher probability of being near $\theta = 0$ than in the opposite direction.

In order to derive now the macroscopic dielectric properties of dilute solutions of dipolar molecules in non-polar liquids, consider an assembly of non-interacting dipoles and let

$$N(\theta, t) \sin\theta \, d\theta$$

be the number of dipoles per unit volume in a range $d\theta$ near θ at time t. The function $N(\theta, t)$ in general varies with time because individual dipoles continuously change their directions. These changes are either due to thermal fluctuations or to the action of the field. The latter, according to 11.19, causes all dipoles to be displaced by an amount $\delta\theta$,

$$\delta\theta = \frac{d\theta}{dt}\delta t = -\frac{\mu f}{\xi}\sin\theta \, \delta t, \qquad (11.20)$$

in a short time interval δt. Therefore within δt seconds a number $N(\theta, t)\sin\theta \, \delta\theta$ will pass through the surface of a cone with angle θ. Owing to the action of the field the rate of change of the number of dipoles within an interval $d\theta$ near θ is, therefore, using 11.20, given by

$$-\frac{d\theta}{\delta t}\frac{\partial}{\partial\theta}\{N(\theta, t)\sin\theta \, \delta\theta\} = \frac{\mu f}{\xi}d\theta\frac{\partial}{\partial\theta}\{N(\theta, t)\sin^2\theta\}.$$

$$(11.21)$$

This equation follows because a number $N(\theta, t)\sin\theta \, \delta\theta$ will leave the range $d\theta$ within δt seconds, whereas the number of dipoles entering it is given by the same expression replacing θ by $\theta - d\theta$.

From 11.21 it follows that the rate of change of the function $N(\theta, t)$ itself due to the action of the field is given by

$$\left(\frac{\partial N(\theta, t)}{\partial t}\right)_f = \frac{\mu f}{\xi} \frac{1}{\sin \theta} \frac{\partial}{\partial \theta} \{N(\theta, t)\sin^2\theta\}. \quad (11.22)$$

In the case of large fluctuations with which we are dealing at present this equation is more suitable than 11.19 to express the effect of the field on dipoles.

Equation 11.22 can be used to find the rate of change of the dipole moment M_f in the field direction, of the assembly of dipoles under the influence of the field. Assuming the dipoles to be rigid,

$$M_f = \mu \int_0^\pi \cos\theta \, N(\theta, t)\sin\theta \, d\theta \quad (11.23)$$

per unit volume. Therefore using 11.22 the rate of change of M_f due to the action of the field is given by

$$\left(\frac{\partial M_f}{\partial t}\right)_f = \mu \int_0^\pi \cos\theta \left(\frac{\partial N(\theta, t)}{\partial t}\right)_f \sin\theta \, d\theta$$

$$= \frac{\mu^2 f}{\xi} \int_0^\pi \cos\theta \frac{\partial}{\partial \theta} \{N(\theta, t)\sin^2\theta\} d\theta,$$

or integrating in parts,

$$\left(\frac{\partial M_f}{\partial t}\right)_f = \frac{\mu^2 f}{\xi} \int_0^\pi \{N(\theta, t)\sin^2\theta\}\sin\theta \, d\theta = \frac{\mu^2 f}{\xi} N_0 \overline{\sin^2\theta}, \quad (11.24)$$

where N_0 is the total number of dipoles per unit volume and $\overline{\sin^2\theta}$ is the average value of $\sin^2\theta$. As discussed above, the field influences the angular distribution of dipoles only very slightly. Therefore the value of $\overline{\sin^2\theta}$ in equilibrium in the absence of a field can be inserted in 11.24, i.e.

$$\overline{\sin^2\theta} = \int_0^\pi \sin^2\theta \sin\theta \, d\theta \bigg/ \int_0^\pi \sin\theta \, d\theta = \tfrac{2}{3}. \quad (11.25)$$

Hence
$$\left(\frac{\partial M_f}{\partial t}\right)_f = \frac{2}{3} \frac{\mu^2 f N_0}{\xi}. \quad (11.26)$$

The total rate of change of M_f contains in addition to 11.26 a term due to thermal motion. This term will tend to restore the equilibrium distribution in the absence of a field, in which case $M_f = 0$. On the assumptions which we have made, namely (i) that there is no interaction between dipoles, (ii) that in a short time interval δt the θ-value of a dipole is altered only very slightly, this second term must be a linear function of the deviation from equilibrium for $f = 0$, i.e. it must be proportional to $-M_f$. Therefore introducing $1/\tau$ as proportionality factor, the total rate of change of M_f is given, using 11.26, by

$$\frac{dM_f}{dt} = -\frac{M_f}{\tau} + \frac{2}{3}\frac{\mu^2 f N_0}{\xi}. \tag{11.27}$$

In view of the linear relationship between M_f and dM_f/dt equation 11.27 leads, of course, to exponential approach to equilibrium and hence, according to § 10, to the Debye equations. This result in fact should always be expected when the following conditions hold:

(a) Absence of interaction between dipoles.

(b) Only one process leading to equilibrium (e.g. either transition over a potential barrier, *or* frictional rotation).

(c) All dipoles can be considered as in equivalent positions, i.e. on an average they all behave in a similar way.

The value of τ can be derived from 11.27 by making use of the equilibrium value of M_f in the presence of a field. Thus, using 6.15,

$$M_f = \frac{\epsilon_s - \epsilon_\infty}{4\pi} f = \frac{\mu^2 N_0 f}{3kT}. \tag{11.28}$$

This expression must be equal to the equilibrium value for M_f resulting from 11.27 with $dM_f/dt = 0$, i.e.

$$M_f = \frac{2}{3}\frac{\tau \mu^2 f N_0}{\xi}. \tag{11.29}$$

Hence comparing 11.28 and 11.29 and using 11.18,

$$\tau = \frac{\xi}{2kT} = \frac{4\pi \eta a^3}{kT}. \tag{11.30}$$

This important formula due to Debye [D2] is often used to discuss the relaxation time of dipolar liquids. It seems appropriate, therefore, to recapitulate the assumptions under which it should hold. First of all the above-mentioned assumptions (a), (b), and (c) require

   (α) dilute solutions of dipolar molecules in a non-polar liquid;
   (β) axially symmetric molecules;
   (γ) isotropy of the liquid—even on an atomic scale in the time average over an interval small compared with τ.

Here (α) follows from (a), and (β) and (γ) are both connected with (b) and (c). All three assumptions are necessary for the Debye equations to hold; but they are not sufficient to obtain the value 11.30 for the relaxation time. This relation, as has been discussed above, is based on the further assumption that the dipolar molecule is fixed fairly rigidly relative to its neighbours so that large jumps of the dipole direction are unlikely. Now the empirical temperature-dependence of the viscosity,

$$\eta \propto e^{H_\eta/kT} \qquad (11.31)$$

suggests that jumps over a potential barrier of height H_η are carried out by the molecules of the liquid in processes connected with viscous flow. Therefore if H is the height of the potential barrier related to jumps of the dipolar molecule, the condition that these jumps happen only very rarely suggests that

(δ) $\qquad\qquad H \gg H_\eta.$

This condition, however, is correct only if the coefficients A (cf. 11.3) have the same order of magnitude for both types of transition. In general it should be expected [$F12, S3$] that low viscosity liquids have small values of H_η and they are more likely, therefore, to satisfy (δ) than are high viscosity liquids.

Formula 11.30 can thus be used to find the temperature dependence of τ which is essentially the same as of η, i.e.

$$\tau \propto e^{H_\eta/kT}, \qquad (11.32)$$

because the other terms vary only slowly with T. It follows that this temperature-dependence is independent of the nature of the dissolved molecules and is a function of the viscosity of the solvent only. To obtain the absolute value of τ as well would require knowledge of the effective radius a which is the radius of a solid sphere having the same frictional constant as the dipolar molecule. No theoretical investigations on the correlation of this radius to the molecular radius normally used have yet been carried out. One should expect, however, that a depends on both solvent and solute and, therefore, cannot be assumed to be a molecular constant. Thus at the present stage of development equation 11.30 cannot be used to obtain absolute values for the relaxation time.

12. Generalizations

The derivation of Debye's relation 11.30—connecting the relaxation time of a dipole molecule with the viscosity of the liquid in which it is dissolved—required the assumption that the dipolar molecule is bound so strongly to the surrounding molecules that large jumps of the dipole direction are very unlikely. This may be true for a number of cases, but others may exist in which the opposite is more likely. A dipolar molecule will then make many jumps over the potential barrier separating it from another dipole direction during the time required for an appreciable change in direction by viscous flow. Clearly this holds for solids where flow may be considered as entirely absent; but sometimes it may also be expected in liquids, and in particular in amorphous substances for which the viscosity is so high that flow is practically negligible. In liquids it might also happen that the process which prevails is different for different kinds of dissolved molecules. A further possibility mentioned by Schallamach [$S3$] is the coexistence of both types of transitions.

As an example of the type of substance exhibiting the second type of behaviour mentioned above we may consider a dilute solution of dipolar molecules in a liquid or in an amorphous solid. In this case, in contrast to § 11, we assume that the dominant process for changing the dipole direction is that involving many large jumps. A single dipole will then behave similarly to a dipole in a crystalline solid; as in 11.3, the probability for a jump over the potential barrier (height H) will be proportional to $\exp(-H/kT)$. In contrast to conditions in crystalline solids, however, the arrangement of the nearest neighbours is not exactly the same for all dipoles, and hence the heights H of the potential barriers will also differ, thus producing different values of the transition probabilities. The dipolar molecules in such a substance may then be classified according to the height H of their respective potential barriers. If the substance has been polarized, and the external field has been removed, the contribution of the dipoles in a small range of energies near H will decay exponentially—as shown in § 11—with a relaxation time τ related to H by equation 11.3. Therefore, instead of H the individual relaxation time τ may be used to classify the molecule. Let $y(\tau)d\tau$ be the contribution to the static dielectric constant of the group of dipoles having individual relaxation times in a range $d\tau$ near τ. Since interaction between dipoles can be considered to be absent (dilute solution), the contributions of the various groups superpose linearly. Their total contribution to the static dielectric constant is, therefore, given by

$$\epsilon_s - \epsilon_\infty = \int_0^\infty y(\tau)\,d\tau. \tag{12.1}$$

The function $y(\tau)$ which describes the distribution of relaxation times will be called the distribution function.

To obtain the complex dielectric constant we first consider the decay function $\alpha(t)$ (cf. §§ 2 and 10). The dipoles with relaxation times in a range $d\tau$ near τ make a contribution to $\alpha(t)$ which is proportional to $\exp(-t/\tau)$ and to $y(\tau)d\tau/\tau$, which corresponds to the coefficient of the exponential term in equation 10.8. Therefore the total contribution of all the dipoles is given by

$$\alpha(t) = \int_0^\infty e^{-t/\tau} y(\tau) \frac{d\tau}{\tau}. \tag{12.2}$$

Using relation 9.1, the complex dielectric constant is now obtained from 12.2,

$$\epsilon(\omega) - \epsilon_\infty = \int_0^\infty \alpha(x) e^{i\omega x}\,dx = \int_0^\infty dx \left[e^{i\omega x} \int_0^\infty \frac{d\tau}{\tau} e^{-x/\tau} y(\tau) \right]$$

$$= \int_0^\infty \frac{d\tau}{\tau} y(\tau) \int_0^\infty dx\, e^{i\omega x - x/\tau}. \tag{12.3}$$

The last step is due to an interchange of x and τ integration. The x-integral is then identical with the integral in 10.14. Therefore

$$\epsilon(\omega)-\epsilon_\infty = \int_0^\infty \frac{y(\tau)\,d\tau}{1-i\omega\tau}, \qquad (12.4)$$

or separating real and imaginary parts according to 2.8 similarly to 10.15 and 10.16,

$$\epsilon_1(\omega)-\epsilon_\infty = \int_0^\infty \frac{y(\tau)\,d\tau}{1+\omega^2\tau^2}, \qquad (12.5)$$

$$\epsilon_2(\omega) = \int_0^\infty \frac{y(\tau)\omega\tau\,d\tau}{1+\omega^2\tau^2}. \qquad (12.6)$$

For a substance of the type described in this section these equations will replace the Debye equations (10.15, 10.16, 10.17), and the latter will not be satisfied. A clearer insight into the meaning of equations 12.5, 12.6 will be obtained by a discussion of the way in which their solutions deviate from those of the Debye formulae.

These deviations are best considered in connexion with the shape of the power loss—frequency curves $\epsilon_2(\omega)$. For a detailed discussion, knowledge of the distribution function $y(\tau)$ is, of course, essential. As a preliminary step, however, we may note that $y(\tau)$ is always positive and that $\epsilon_2(\omega)$ therefore consists of a superposition of Debye curves $\omega\tau/(1+\omega^2\tau^2)$ (cf. Fig. 13) with different positions of their respective maxima. The resulting curve $\epsilon_2(\omega)$, therefore—supposing it has a single maximum—has a larger half-width than that of a single Debye curve whose maximum coincides with that of $\epsilon_2(\omega)$. As a simple example [F7] consider a model in which each molecule has two equilibrium positions with opposite dipole directions, and with equal energy in the ground level, as in the case of disordered solids; but in contrast to the latter the potential barrier between the two positions has a different height for each molecule. It will be assumed that the heights H of the potential barriers are equally distributed over a range between H_0 and H_0+v_0, i.e.

$$H = H_0+v, \qquad 0 \leqslant v \leqslant v_0. \qquad (12.7)$$

Thus if N_0 is the total number of dipoles per unit volume,

$$N_0 \frac{dv}{v_0} \tag{12.8}$$

is the fraction with H-values in a range dv near H_0+v.

For dilute solutions interaction between dipoles can be neglected. As shown in §§ 4 and 6, the contribution of a dipolar molecule to ϵ_s is then independent of H, and therefore is the same for all molecules, i.e. $(\epsilon_s-\epsilon_\infty)/N_0$ per molecule. The individual relaxation time τ, however, depends on H according to 11.3 and 11.8. Hence, using 12.7 and considering A as constant,

$$\tau = \tau_0 e^{v/kT}, \qquad \tau_0 = \frac{\pi}{2\omega_a} A e^{H_0/kT}. \tag{12.9}$$

The individual relaxation times τ therefore cover the range

$$\tau_0 \leqslant \tau \leqslant \tau_1 \quad \text{where} \quad \tau_1 = \tau_0 e^{v_0/kT}. \tag{12.10}$$

To determine the distribution function $y(\tau)$ we note that $y(\tau) = 0$ outside the range 12.10. Now consider τ as a function of v so that from 12.1, using 12.9,

$$\epsilon_s-\epsilon_\infty = \int_{\tau_0}^{\tau_1} y(\tau)\, d\tau = \int_0^{v_0} y(\tau)\frac{d\tau}{dv}\, dv = \frac{1}{kT}\int_0^{v_0} y(\tau)\tau(v)\, dv. \tag{12.11}$$

This means that $y(\tau)\tau\, dv/kT$ is the contribution of the molecules in the range dv to the static dielectric constant. On the other hand, we have just seen that this contribution per molecule is $(\epsilon_s-\epsilon_\infty)/N_0$ independent of v. Therefore, since 12.8 represents the number of molecules in dv,

$$\frac{y(\tau)\tau\, dv}{kT} = \frac{\epsilon_s-\epsilon_\infty}{N_0} N_0 \frac{dv}{v_0}, \tag{12.12}$$

or

$$\left.\begin{array}{l} y(\tau) = (\epsilon_s-\epsilon_\infty)\dfrac{kT}{v_0}\dfrac{1}{\tau}, \quad \text{if} \quad \tau_0 \leqslant \tau \leqslant \tau_1 = \tau_0 e^{v_0/kT} \\[6pt] y(\tau) = 0, \qquad\qquad\qquad \text{if} \quad \tau < \tau_0 \ \text{and} \ \tau > \tau_1 \end{array}\right\} \tag{12.13}$$

The above distribution function—corresponding to an equal distribution of potential barriers over a range v_0—depends on temperature. In particular its relative width $(\tau_1-\tau_0)/\tau_0$ decreases with temperature since

$$\frac{\tau_1-\tau_0}{\tau_0} = e^{v_0/kT}-1. \qquad (12.14)$$

The dielectric constants ϵ_1 and ϵ_2 can now be obtained by

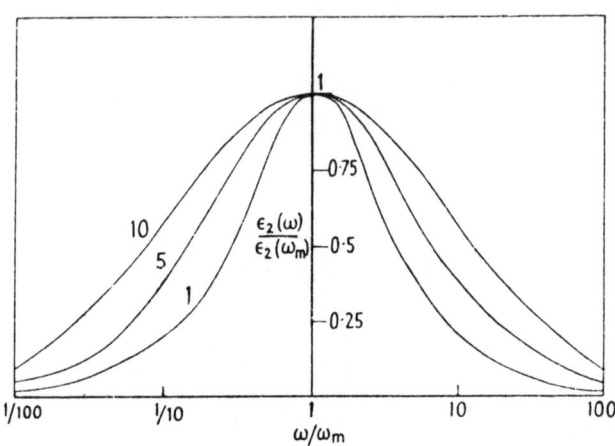

Fig. 14. Dependence of dielectric loss $\epsilon_2(\omega)$ on angular frequency ω according to equation 12.19 for the three values 1, 5, and 10 of the parameter $\sqrt{(\tau_1/\tau_0)}$. They correspond to a range of heights of potential barriers of width $v_0 = kT\log\tau_1/\tau_0$. $\epsilon_2(\omega)/\epsilon_2(\omega_m)$ is plotted against ω/ω_m on a logarithmic scale; ω_m represents the value of ω where ϵ_2 reaches its maximum.

inserting $y(\tau)$ from 12.13 into 12.5 and 12.6. All integrations can be carried out in an elementary manner leading to

$$\epsilon_1(\omega)-\epsilon_\infty = (\epsilon_s-\epsilon_\infty)\left(1-\frac{kT}{2v_0}\log\frac{1+\omega^2\tau_0^2 e^{2v_0/kT}}{1+\omega^2\tau_0^2}\right), \qquad (12.15)$$

$$\epsilon_2(\omega) = (\epsilon_s-\epsilon_\infty)\frac{kT}{v_0}(\tan^{-1}(\omega\tau_0 e^{v_0/kT})-\tan^{-1}\omega\tau_0). \qquad (12.16)$$

These equations replace the Debye formulae 10.15 and 10.16 for our present model. In this case $\epsilon_1(\omega)$ and $\epsilon_2(\omega)$, considered as functions of frequency, depend on two parameters—a relaxation time τ_0, and the factor v_0/kT determining the width of the range of relaxation times by 12.14. The Debye formulae are obtained for $v_0/kT = 0$; they contain, therefore, one parameter only. The shape of ϵ_2 can be seen from Fig. 14, where

$\epsilon_2(\omega)/\epsilon_2(\omega_m)$ is plotted against ω/ω_m, ω_m being the frequency at which ϵ_2 has its maximum.

It can be seen that with increasing values of the second parameter v_0/kT the curves become more and more flattened out. The position of the maximum is obtained from 12.16, using

$$\frac{\partial \epsilon_2}{\partial \omega} = 0 \quad \text{if} \quad \omega = \omega_m.$$

This leads to
$$\omega_m = \frac{1}{\tau_0} e^{-\frac{1}{2}v_0/kT} = \frac{1}{\sqrt{(\tau_0 \tau_1)}}, \qquad (12.17)$$

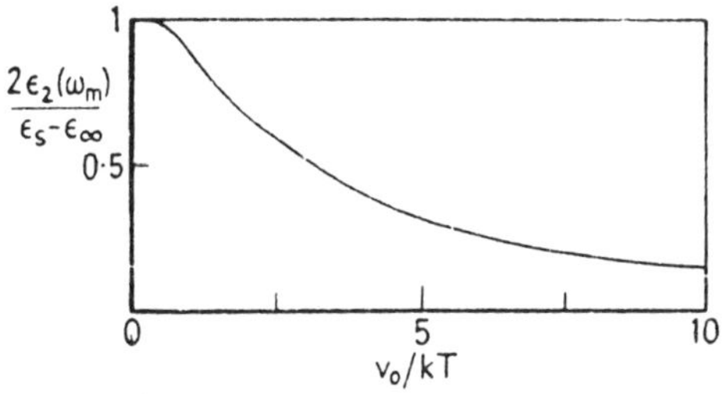

Fig. 15. Dependence of the maximum value of the loss, $\epsilon_2(\omega_m)$, on the width v_0 of the range of heights of potential barriers.

which for $v_0 = 0$, of course, becomes identical with the corresponding value 10.22 for the Debye formulae. Inserting 12.17 into 12.16 yields the maximum value of ϵ_2,

$$\epsilon_2(\omega_m) = (\epsilon_s - \epsilon_\infty) \frac{kT}{v_0} [\tan^{-1} e^{\frac{1}{2}v_0/kT} - \tan^{-1} e^{-\frac{1}{2}v_0/kT}]. \qquad (12.18)$$

Again for $v_0 = 0$ this quantity becomes equal to 10.23, but it decreases with increasing v_0/kT, as shown in Fig. 15.

With the help of 12.14, 12.17, and 12.18 equation 12.16 can be written in the form

$$\frac{\epsilon_2(\omega)}{\epsilon_2(\omega_m)} = \frac{\tan^{-1}\frac{\omega}{\omega_m}\sqrt{\left(\frac{\tau_1}{\tau_0}\right)} - \tan^{-1}\frac{\omega}{\omega_m}\sqrt{\left(\frac{\tau_0}{\tau_1}\right)}}{\tan^{-1}\sqrt{\left(\frac{\tau_1}{\tau_0}\right)} - \tan^{-1}\sqrt{\left(\frac{\tau_0}{\tau_1}\right)}}, \qquad (12.19)$$

which shows clearly the dependence of $\epsilon_2(\omega)$ on the parameter $\tau_0/\tau_1 \leqslant 1$. For $\tau_0/\tau_1 = 1$, equation 12.19 is identical with the corresponding equation of § 10, and the smaller τ_0/τ_1 is, that is the larger v_0, the larger is the deviation from a Debye curve.

The above equations are based on the assumption of equal distribution of the height of potential barriers over a range v_0 leading to a range of relaxation times given by equation 12.14 and to the distribution function 12.13. This may seem to be rather a specialized assumption. It can be shown quite readily, however, that the behaviour of $\epsilon_2(\omega)$ in the main absorption range is altered but little, if the distribution function 12.13 is replaced by any other smooth function which is large between τ_0 and τ_1 and small outside this range. Quite generally, therefore, in the neighbourhood of the maximum of $\epsilon_2(\omega)$ this function can be considered to be determined by two parameters, namely the angular frequency ω_m at which the maximum occurs, and the width of the range of relaxation times $\tau_1 - \tau_0$. It should be remembered that the absolute value of $\epsilon_2(\omega)$ is then determined by $\epsilon_s - \epsilon_\infty$ with the help of the phenomenological relation 2.18.

In contrast to the behaviour of $\epsilon_2(\omega)$ in the main absorption region the behaviour outside this region may be very sensitive to small changes of the distribution function $y(\tau)$ for τ-values outside the range $\tau_0 < \tau < \tau_1$, which contains the bulk of the relaxation times. For the latter, though very numerous, make only small contributions to $\epsilon_2(\omega)$ if $\omega\tau_0 \ll 1$ or $\omega\tau_1 \gg 1$. These may well be smaller than the contributions of a few relaxation times near $\tau \sim 1/\omega$ which make their maximum contribution at the frequencies outside of the main absorption band.

For many substances the value of $\epsilon_2(\omega)$ does not approach zero outside the main absorption range, but remains at a very small value which changes very slowly with frequency [$G2$]. This may be explained if a sufficiently wide range of relaxation times is used. Garton [$G1$] has suggested that this residual absorption may be due to the temporary formation of additional equilibrium positions of the molecules due to thermal fluctuations. The probability for the occurrence of such a potential well of depth w in a range dw (cf. Fig. 16) is assumed to be proportional to

$$e^{-w/kT}\,dw. \tag{12.20}$$

This is a plausible assumption, though no proof for it has been given. We shall proceed to show that such an assumption does, in fact, lead to a value of $\epsilon_2(\omega)$ which is substantially independent of frequency for frequencies well outside of the main absorption range.

Let H be the height of the potential barrier measured from a

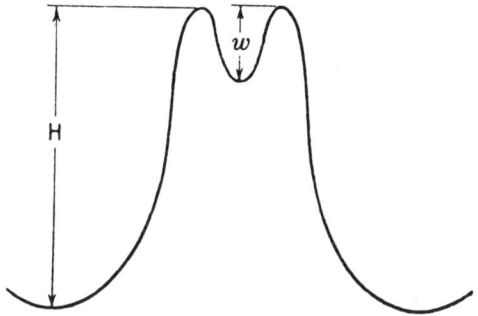

Fig. 16. Potential well of depth w due to thermal fluctuations.

normal (permanent) position and consider a molecule near which a temporary position of depth w has been formed. According to 11.3 the time it spends in the temporary or in the permanent position is given by

$$\tau = Be^{w/kT} \quad \text{and} \quad \tau_0 = B_0 e^{H/kT} \qquad (12.21)$$

respectively on the further assumption that the temporary position exists for a time which is long compared with $\tau+\tau_0$. The coefficients B and B_0 are considered to be independent of w. Assuming $\tau \ll \tau_0$, the relative probability of finding this molecule in the temporary position is given by $\tau/(\tau+\tau_0) \simeq \tau/\tau_0$. The average number of molecules in such positions, using 12.20 and 12.21, is thus proportional to

$$\frac{\tau}{\tau_0} e^{-w/kT} dw = \frac{\tau}{\tau_0} \frac{B}{\tau} \frac{dw}{d\tau} d\tau = \frac{B}{\tau_0} kT \frac{d\tau}{\tau} \propto \frac{d\tau}{\tau}. \qquad (12.22)$$

Each molecule makes a contribution to $\epsilon_2(\omega)$ proportional to $\omega\tau/(1+\omega^2\tau^2)$. Also since $w > 0$ the smallest value that τ can take is $\tau = B$. For large values near $\tau = \tau_0$ the developments become invalid, and τ_0 will be used as the upper limit for τ; this should not be an essential limitation, however, because near $\tau = \tau_0$ the main contribution to $\epsilon_2(\omega)$ is due to the transitions between permanent positions which are much more numerous. Thus

$$\epsilon_2(\omega) \propto \int_B^{\tau_0} \frac{\omega\tau}{1+\omega^2\tau^2} \frac{d\tau}{\tau} = \tan^{-1}\omega\tau_0 - \tan^{-1}B\omega \quad (12.23)$$

for frequencies higher than the main absorption range, $\omega\tau_0 \gg 1$. If, in particular,

$$1/\tau_0 \ll \omega \ll 1/B, \quad (12.24)$$

then $\epsilon_2(\omega)$ is practically independent of ω. Here $1/B$ will probably be a very high frequency.

The preceding discussions were all based on the assumption that interaction between dipoles can be neglected. It seems likely that the influence of interaction will always tend to broaden the Debye absorption curve, because the energy of a molecule in an equilibrium position may have a whole range of values depending on the positions of its neighbours. One would thus expect the heights of the potential barriers to vary accordingly. In the present case, however, the time which a molecule spends in a given equilibrium position is of the same order as the time during which the height of the potential barrier remains constant. If an attempt is made to include interaction under these conditions, great mathematical difficulties are encountered and no solution for this case has yet been obtained.

13. Resonance absorption

In § 9 it was pointed out that two types of power loss should be expected to occur in dielectrics: (i) The loss due to displacement of charges bound elastically to an equilibrium position. Such charges have a proper frequency of oscillation, say $\omega_0/2\pi$, and the power loss (and hence ϵ_2) will be expected to have a maximum near this frequency (the resonance frequency). (ii) The loss due to transitions of charges or dipoles between equilibrium positions separated by a potential barrier. Such transitions are described by a relaxation time τ, and in this case the power loss will have a maximum near a frequency $1/2\pi\tau$ as discussed in §§ 10–12. This frequency is usually strongly dependent on temperature, in contrast to the resonance frequency in case (i). The chief aim of the present section is to derive expressions for the complex dielectric constant when there is resonance absorption (case (i) above).

It is our intention to consider only the simplest case, and it will be seen presently that the required formula can be derived from very general considerations without choosing a particular model. We shall employ the same method that has been used in § 10 to derive the Debye equations. The polarization of the substances considered there was assumed to approach equilibrium exponentially with time. This cannot be expected to hold in the present case of elastic displacement. One would rather expect damped oscillations with frequency $\omega_0/2\pi$ about the equilibrium polarization, as has already been discussed in § 9. Therefore instead of 10.1 we shall now assume a decay function

$$\alpha(t) = \gamma e^{-t/\tau}\cos(\omega_0 t + \psi), \qquad (13.1)$$

and then, making use of equation 9.1, derive an expression for the complex dielectric constant ϵ. Equation 13.1 contains two constants γ and ψ which will be determined below.

The reader is reminded that $\alpha(t)$ is a macroscopic quantity referring to the behaviour of the polarization of the whole substance. It would be wrong to describe the amplitude of the oscillations of a single molecule (in the absence of a field) by a function similar to 13.1, because in equilibrium this would lead to vanishing amplitude and thus would not take account of the thermal motion.

By inserting 13.1 into 9.1 we now find for the complex dielectric constant ϵ,

$$\begin{aligned}
\epsilon - \epsilon_\infty &= \gamma \int_0^\infty e^{-x/\tau}\cos(\omega_0 x + \psi) e^{i\omega x} dx \\
&= \frac{\gamma\tau}{2}\left(\frac{e^{i\psi}}{1-i(\omega_0+\omega)\tau} + \frac{e^{-i\psi}}{1+i(\omega_0-\omega)\tau}\right) \\
&= \frac{\gamma\tau}{2}\cos\psi\left(\frac{1+i\tan\psi}{1-i(\omega_0+\omega)\tau} + \frac{1-i\tan\psi}{1+i(\omega_0-\omega)\tau}\right). \qquad (13.2)
\end{aligned}$$

As the next step, the two constants γ and ψ will be determined by considering the limiting cases of very low and very high frequencies. Low frequencies, $\omega \ll \omega_0$, are considered to be outside the main range of absorption, i.e. in a region in which ϵ_2 is very small and, therefore, the dielectric constant is practically real. In this range ϵ will approach a constant value, say

$$\epsilon = \epsilon_\infty + \Delta\epsilon, \qquad \omega \ll \omega_0, \qquad (13.3)$$

where $\Delta\epsilon$ is a real quantity. This value of the dielectric constant will hold down to frequencies at which another absorption range starts. In the absence of further absorption regions 13.3 represents the static dielectric constant. The high-frequency dielectric constant ϵ_∞, on the other hand, is the value of ϵ at the high-frequency side of the absorption peak, again in a region in which ϵ_2 is very small.

Comparing 13.3 with equation 13.2 and neglecting ω compared with ω_0, we find

$$\Delta\epsilon = \gamma\tau \frac{\cos\psi - \omega_0\tau\sin\psi}{1+\omega_0^2\tau^2} = \gamma\tau\cos\psi \frac{1-\omega_0\tau\tan\psi}{1+\omega_0^2\tau^2} \qquad (13.4)$$

as a first condition for the determination of γ and ψ.

The second condition is of a more subtle nature and concerns the behaviour of ϵ at angular frequencies $\omega \gg \omega_0$. At these frequencies ω_0 can be neglected compared with ω, so that 13.2 becomes

$$\epsilon - \epsilon_\infty = \gamma\tau \frac{\cos\psi}{1-i\omega\tau}, \qquad \omega \gg \omega_0. \qquad (13.5)$$

On the other hand, for $\omega \gg \omega_0$ the influence of the restoring force should be negligible during one period $1/\omega$. Therefore a behaviour according to the Debye theory should be expected with a static dielectric constant given by 13.3. Thus by replacing $\epsilon_s - \epsilon_\infty$ by $\Delta\epsilon$ in 10.13,

$$\epsilon - \epsilon_\infty = \frac{\Delta\epsilon}{1-i\omega\tau}, \qquad \omega \gg \omega_0. \qquad (13.6)$$

Combining 13.5 and 13.6 we find

$$\Delta\epsilon = \gamma\tau\cos\psi \qquad (13.7)$$

as the second condition for the determination of γ and ψ.

From 13.7 and 13.4 we now conclude that
$$\omega_0 \tau = -\tan\psi. \tag{13.8}$$
The insertion of this expression and 13.7 into 13.2 finally leads to† (cf. references $V5$ and $F9$)
$$\epsilon - \epsilon_\infty = \tfrac{1}{2}\Delta\epsilon\left(\frac{1-i\omega_0\tau}{1-i(\omega_0+\omega)\tau} + \frac{1+i\omega_0\tau}{1+i(\omega_0-\omega)\tau}\right), \tag{13.9}$$
or separating real and imaginary parts according to 2.8,
$$\epsilon_1(\omega) - \epsilon_\infty = \tfrac{1}{2}\Delta\epsilon\left(\frac{1+\omega_0(\omega+\omega_0)\tau^2}{1+(\omega+\omega_0)^2\tau^2} + \frac{1-\omega_0(\omega-\omega_0)\tau^2}{1+(\omega-\omega_0)^2\tau^2}\right) \tag{13.10}$$
and
$$\epsilon_2(\omega) = \tfrac{1}{2}\Delta\epsilon\left(\frac{\omega\tau}{1+(\omega+\omega_0)^2\tau^2} + \frac{\omega\tau}{1+(\omega-\omega_0)^2\tau^2}\right). \tag{13.11}$$

These equations represent the dielectric constants in the simplest case of resonance absorption. In many practical cases $\Delta\epsilon$ is a small quantity so that ϵ_1 is approximately equal to ϵ_∞. In these cases the loss angle according to 2.5 is given by
$$\tan\phi = \frac{\epsilon_2}{\epsilon_1} \simeq \frac{\epsilon_2}{\epsilon_\infty}, \quad \text{if } \Delta\epsilon \ll 1. \tag{13.12}$$
If $\Delta\epsilon$ is small it will be difficult to determine this quantity experimentally with reasonable accuracy by direct measurement. Following 2.18, however, the relation
$$\Delta\epsilon = \frac{2}{\pi}\int_0^\infty \epsilon_2(\omega)\frac{d\omega}{\omega} = \frac{2}{\pi}\int \epsilon_2 \, d(\log\omega) \tag{13.13}$$
can be used instead. This equation holds for the present case, as can be checked by inserting ϵ_2 from 13.11.

We shall now proceed to discuss power loss ($\propto \epsilon_2$) in more detail. At constant temperature ϵ_2 as a function of ω has a maximum when $\omega = \omega_m$,
$$\omega_m = \frac{1}{\tau}\sqrt{(1+\omega_0^2\tau^2)}, \tag{13.14}$$
as can be found from the condition $\partial\epsilon_2/\partial\omega = 0$. The maximum value of ϵ_2 is then given by (inserting 13.14 into 13.11)
$$\epsilon_2(\omega_m) = \tfrac{1}{2}\Delta\epsilon\sqrt{(1+\omega_0^2\tau^2)} = \tfrac{1}{2}\Delta\epsilon\,\omega_m\tau. \tag{13.15}$$

† Another derivation of 13.9 is given in the Appendix A 4.

Consider now that the time of relaxation τ depends on temperature and may be expected to decrease with increasing temperature.

Then
$$\omega_m \simeq 1/\tau \quad \text{and} \quad \epsilon_2(\omega_m) \simeq \tfrac{1}{2}\Delta\epsilon \quad \text{if} \quad \omega_0\tau \ll 1, \quad (13.16)$$
i.e. at very high temperatures. But
$$\omega_m \simeq \omega_0 \quad \text{and} \quad \epsilon_2(\omega_m) \simeq \tfrac{1}{2}\Delta\epsilon\,\omega_0\tau \quad \text{if} \quad \omega_0\tau \gg 1, \quad (13.17)$$
i.e. at lower temperatures. Comparison with 10.23 shows that case 13.16 leads to behaviour similar to that for Debye absorption. Without detailed consideration of a special model it is impossible, however, to say whether this region can be reached at temperatures corresponding to the solid state. The low-temperature limiting case 13.17 leads to typical resonance absorption. In contrast to Debye absorption, the angular frequency ω_0 of maximum absorption is independent of temperature, but the resonance peak becomes narrower and higher the lower the temperature, i.e. the larger $\omega_0\tau$. Figures 17 and 18 give a comparison of Debye absorption and resonance absorption for various temperatures.

A number of models have been considered in greater detail, and equations 13.9–13.11 have been found to hold. These derivations are much more complicated than the one given above, but they may be considered to be more rigid (cf. Appendix A 4). Transitions between rotational levels of gases have been considered by van Vleck and Weisskopf [*V5*] and by van Vleck [*V4*]. In gases the relaxation process is due to collisions between molecules. Therefore τ will be expected to decrease with increasing pressure and with increasing temperature. At temperatures or pressures at which $\omega_0\tau \ll 1$ the absorption peak according to 13.16 is near the frequency $\omega_m = 1/\tau$. In this region, therefore, $\omega \gg \omega_0$, so that ω_0 can be neglected.

In the main absorption region, therefore, the Debye equations are fulfilled whenever $\omega_0\tau \ll 1$. At small pressures, however, $\omega_0\tau \gg 1$ and the main absorption peak according to 13.17 will be found near the resonance frequency ω_0.

Resonance absorption may also be of importance in solids

containing dipolar molecules with large moments of inertia
[*F9, H3, S11*]. The resonance in this case is due to rotational

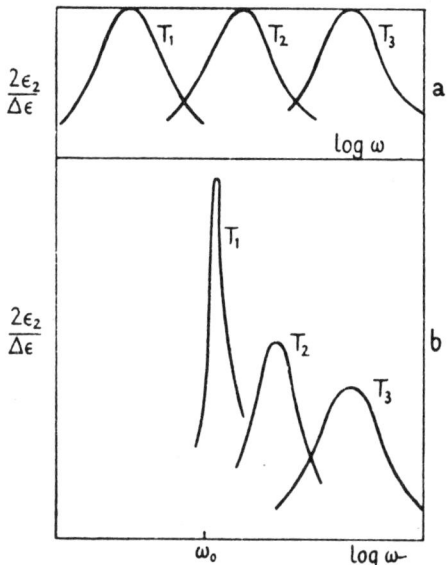

Fig. 17. Frequency dependence of dielectric loss for three different temperatures $T_1 < T_2 < T_3$ (schematically). (a) Debye case; (b) resonance case.

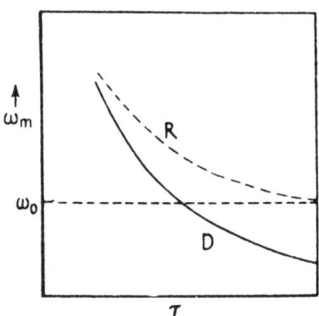

Fig. 18. Dependence of the angular frequency ω_m at which dielectric loss has its maximum on the relaxation time. D refers to the Debye case, R to the resonance case.

oscillations of the molecules about their equilibrium positions. The behaviour of the relaxation time in this case, however, has not yet been investigated.

CHAPTER IV

APPLICATIONS

THE present chapter is designed to show by examples how the general theory developed in previous chapters can be applied in the discussion of the properties of dielectric materials. It is the intention to cover some important types of materials (e.g. non-polar solids, ionic crystals, dipolar liquids, etc.); this should, however, not be considered as an attempt to discuss systematically the properties of dielectric materials, but rather as a selection of typical examples.

14. Structure and dielectric properties

Atoms

In the present section we shall commence by discussing the structure and the dielectric properties of atoms and molecules leading up to a classification of dielectric materials according to their structure. An atom, as the reader will know, consists of a positively charged nucleus surrounded by electrons which just compensate its charge and whose combined mass is very small compared with that of the nucleus. By adding or removing one or several electrons one obtains negative or positive ions. In a stationary state the electric dipole moment of an atom (or of an ion relative to its nucleus) vanishes.

A very general quantum mechanical theorem says that under the influence of external (static or alternating) electric fields the electrons of an atom behave like an assembly of classical harmonic oscillators whose frequencies and other properties can be specified, but will not be required by us. Usually the frequencies are equal to or higher than those of visible light, and, therefore, are much higher than the frequencies of electric fields contemplated in the present book. Under the influence of an external electric field f an electric dipole m_o will, therefore, be induced which has all the characteristics of a dipole produced by elastic displacement of electric charges (electrons) as discussed in case (i) of §§ 4 and 9. In the ranges of field-strengths and frequencies

to be considered here, the strength of the induced dipole m_o is proportional to the field-strength f, but independent of the frequency, and it shows no phase-shift relative to the field. These properties of an atom can be described by a constant polarizability α_o,
$$\mathbf{m}_o = \alpha_o \mathbf{f}. \qquad (14.1)$$

The resulting polarization will be called optical polarization, to indicate that the characteristic frequencies are usually in the optical region. Often it is also described as electronic polarization because it is due to the displacement of the electrons; this description may, however, be misleading as will be seen presently.

Molecules

Consider now a diatomic molecule AB composed of two atoms A and B. In view of the interaction between the atoms—leading to the chemical bond—the distribution of electrons in the molecule is different from a mere superposition of the electron distributions of the free atoms A and B. The distribution should, however, be axially symmetrical around the line joining the nuclei of the two atoms. It must, therefore, be expected that a diatomic molecule will possess a dipole moment in the direction A–B. An exception occurs in the case of two equal atoms, $A = B$, where the dipole moment must vanish for reasons of symmetry. Thus, for instance, the molecules HCl or CO have a dipole moment in contrast to H_2, O_2, or Cl_2. The magnitude of the dipole moment gives valuable information about the distribution of electrons. Thus the large dipole moment of HCl ($\mu \sim 1 \times 10^{-18}$ e.s.u. = 1 Debye unit) compared with that of CO ($\mu \sim 0.1$ Debye units) gives some justification for considering the HCl molecule (but not the CO molecule) as composed of a positive and a negative ion, $H^+ + Cl^-$.

Under the influence of an external field $\mathbf{f}$ an average dipole moment $\mathbf{m}$ is induced in a molecule which, according to §§ 4 and 9, can be considered as composed of contributions of two types: (i) those due to elastic displacement of charges, and (ii) those due to a change in the average orientation of the permanent dipole of the molecule. The former again are composed of the contributions of the various normal modes. In view of the great difference in the mass of electrons and nuclei there is one group of normal modes which is practically entirely due to displacement of electrons relative to the nuclei. Their proper frequencies are usually in the optical region, visible or ultra-violet. The proper frequencies of the normal modes connected with oscillations of the nuclei, on the other hand, are usually in the infra-red region. They must contain displacements of electrons relative to their respective nuclei as well, because of the variation of the interaction between electrons and nuclei with the distance between the latter. Thus

$$\mathbf{m} = \mathbf{m}_o + \mathbf{m}_{ir} + \mathbf{m}_d, \qquad (14.2)$$

where the first two terms are due to elastic displacements with proper frequencies in the optical and in the infra-red region, and $\mathbf{m}_d$ is the moment due to dipole orientation. The quantities $\mathbf{m}_o$ and $\mathbf{m}_{ir}$ are often denoted as electronic and atomic polarization, respectively, which may be considered as somewhat misleading because the latter also contains terms due to electronic displacements. Since the frequencies contemplated in this book are much smaller than all proper frequencies, the relations

$$\mathbf{m}_o = \alpha_o \mathbf{f}, \qquad \mathbf{m}_{ir} = \alpha_{ir} \mathbf{f} \qquad (14.3)$$

should hold with constant values of the polarizabilities α_o and α_{ir}, independent of the frequency of the field or of temperature.

Polyatomic molecules behave in a similar way to diatomic molecules in that the average dipole moment induced by an external field can be considered as a superposition of three terms according to 14.2: two terms representing the elastic displacements with proper frequencies in the optical and infra-red regions respectively, and a dipolar contribution. The former can be obtained from the optical and the infra-red polarizabilities α_o and α_{ir} according to 14.3; the value of the latter depends on the dipole moment of the molecule (e.g. § 6). The determination of the dipole moment is of great interest in investigations on the structure of molecules. For instance, the fact that CO_2 does not possess a permanent dipole moment leads to the conclusion that the three atoms must be arranged in a straight line with the carbon atom half-way between the two oxygen atoms. In contrast to this the H_2O molecule has a permanent dipole moment, which means that H_2O is a triangular molecule.

A great number of interesting applications to stereochemistry have been made, but a detailed discussion of this subject is beyond the scope of the present book. The reader will find examples and reference to the literature in the fairly recent book by Le Fevre [*L2*] or in the article by Sutton [*S10*]. The earlier developments are to be found in the books by Debye [*D2*] and by Smyth [*S8*].

Two qualitative rules concerning large molecules should be mentioned. Such molecules often contain a number of dipolar molecular groups such as the hydroxyl group O—H, or the ketone group C=O. To a fair approximation these groups contribute the same dipole moment in whatever molecule they occur. The total moment of such a molecule is then the vector sum of the moments of all groups present in the molecule. Clearly this rule cannot hold exactly for several reasons; one of these is the influence of the interaction between the various groups on their individual moments.

The second rule refers to the optical polarizability of large molecules. As mentioned above, optical polarization is due to displacement of electrons on the assumption that all nuclei are kept in fixed positions. Clearly the largest contributions will arise from the electrons with the smallest binding energies, i.e. from valency electrons. The behaviour of these electrons is different in a compound than it is in the atomic state of the atoms contained in the compound. Thus the optical polarizability of CO_2 is different from the sum of the polarizabilities of a carbon and two oxygen atoms. On the other hand, the distribution of electrons in certain molecular groups is to a fair approximation independent of the type of molecule in which they occur, as in the above case of the dipole moment of such groups. In applying this rule it should be remembered that frequently the polarizability of such a group depends on direction; for the ketone group C=O, for instance, the polarizability in the direction of the line connecting the two atoms would be different from that in a direction perpendicular to it.

Classification of dielectrics

The division of polarization into three types which was introduced above is of a very general nature. It leads at once to the following division of dielectric materials into three classes:

(i) Non-polar substances showing optical polarization only.

(ii) Polar substances having optical as well as infra-red polarization.

(iii) Dipolar substances which in addition show also polarization due to dipolar orientation.

In the first class of materials, non-polar ones, an electrical field produces elastic displacement of electrons only. This class contains all dielectrics consisting of a single type of atom, whether they form gases, liquids, or solids. Examples are diamond, oxygen (solid, liquid, or as vapour), the inert gases, and many others. The dielectrics of the second class, polar materials, are capable of infra-red polarization as well as optical polarization. Substances of this type may contain dipolar groups of atoms, but these groups must show only elastic displacement. If, on the other hand, there are several equilibrium positions for the dipole, then the substance belongs to the third class of materials.

The second class contains, first of all, substances consisting of molecules whose total dipole moment vanishes, though they contain dipolar groups of atoms. Examples are CO_2, paraffins CH_3—$(CH_2)_n$—CH_3 (cf. § 15), benzene C_6H_6, carbon tetrachloride CCl_4, and many others in the solid, liquid, or vapour phase. In most of these substances the infra-red polarizability is only a small fraction of the optical polarizability. From a practical point of view, therefore, their behaviour is very similar to that of non-polar substances. At sufficiently low temperatures many solids consisting of dipolar molecules fall into the same category because the dipoles in these solids freeze in, i.e. the thermal energy is insufficient to turn them in a reasonable time into other equilibrium positions. The most representative substances of the polar type are, however, ionic crystals which may show very large infra-red polarizability. Examples are rock-salt

NaCl, other alkali halide crystals, TiO_2 crystals, and most other crystals of salts. In contrast to molecular lattices, which have a whole molecule at each lattice-point, ionic lattices contain one ion at each lattice-point. Thus rock-salt, for instance, forms a simple cubic lattice of Na^+ and Cl^- ions (cf. § 18). In contrast to other dielectrics, most salts on melting become (ionic) conductors.

All materials consisting of dipolar molecules belong to the third class, except in the very low temperature range mentioned above where the dipoles are frozen in. Often in solids this process starts just below a critical temperature at which the substance undergoes a phase transition. In nearly all of these materials the turning of a dipole into another equilibrium position is connected with a turning of the whole molecule (e.g. for ketones, § 17). Exceptions are ice and some other crystals where a turning of the direction of a dipole may be obtained by a transfer of a H^+ ion from one equilibrium position to another.

15. Non-polar substances

The simplest type of dielectric substances show elastic displacement of electrons only, and in the classification of § 14 are designated as non-polar. For such substances the lowest frequency ν_0 at which appreciable absorption occurs is usually in the visible or in the ultra-violet region. For all frequencies ν which are less than ν_0 by a sufficient amount, the dielectric constant should be independent of frequency. Thus for $\nu \ll \nu_0$ the dielectric constant ϵ should be equal to the static dielectric constant ϵ_s and should satisfy the Maxwell relation $\epsilon = n^2$. That is,

$$\epsilon_s = n^2 \qquad (15.1)$$

should hold between the static dielectric constant and the refractive index n at frequencies $\nu \ll \nu_0$. Whitehead and Hackett [W6] have recently checked this relation on diamond and found that it holds within the range of accuracy of the measurements. They measured the dielectric constant at frequencies between 500 and 3,000 cycles per second and obtained the value $5\cdot 68 \pm 0\cdot 03$. The refractive index n was obtained by extrapolating measurements at various wave-lengths in the optical region to long waves where n is independent of wave-length. The accuracy of this procedure can be seen from Fig. 19 and leads to the value $n^2 = 5\cdot 66$. This agreement means that absorption of electric waves at frequencies between 3,000 cycles per second and optical frequencies must be so weak that (cf. 2.18) $(2/\pi)\int \epsilon_2 d\nu/\nu < 0\cdot 03$ if integrated over this range. Actually diamond absorbs at infra-red frequencies, but the reason for this absorption has not

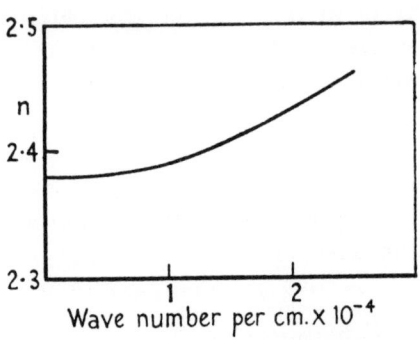

Fig. 19. Dependence of the refractive index n of diamond on the wave number according to measurements collected by Whitehead and Hackett [$W6$].

yet been found. The above condition, by the way, need not imply that ϵ_2 is small provided the absorption bands are sufficiently narrow.

Gases of non-polar molecules fulfil fairly closely the conditions required for the validity of the Clausius–Mossotti formula, (cf. §§ 6 and 8 and Appendix A 3). These conditions require (i) elastic displacement only, (ii) absence of non-dipolar (short-range) interaction between molecules, (iii) isotropy of the polarizability of a molecule, (iv) isotropy or cubic symmetry in the arrangement of the molecules. Condition (i) is fulfilled for all non-polar molecules; (ii) holds so long as the distance between molecules is sufficiently large; (iii) holds for spherical molecules only, and (iv) holds always in gases. Hence deviations may be expected for non-spherical molecules (condition (iii) violated) and at very high pressure (condition (ii) violated). All conditions are thus fulfilled for the rare gases. The main check of the

Clausius–Mossotti formula consists in measurement of the dielectric constant of a gas as a function of its density. For, since the number N_0 of molecules per c.c. is obtained with the help of the Avogadro number ($6{\cdot}02\times 10^{23}$),

$$N_0 = \frac{d}{W} 6{\cdot}02\times 10^{23} \qquad (15.2)$$

(d = density, W = molecular weight).

It follows with 6.34 and 6.32 that

$$\frac{\epsilon_s - 1}{\epsilon_s + 2} = 2{\cdot}52\times 10^{24} \frac{d}{W}\alpha. \qquad (15.3)$$

Often the quantity

$$p = 2{\cdot}52\times 10^{24}\alpha \qquad (15.4)$$

is defined as molecular polarizability. Using 15.3 and the Maxwell relation (cf. 15.1),

$$p = \frac{W}{d}\frac{\epsilon-1}{\epsilon+2} \qquad (15.5)$$

should be a constant independent of density, temperature, or frequency. For the rare gases the following values for p have been found:

	He	Ne	Ar	Kr	X	
$p =$	0·5	1·0	4·2	6·3	10	c.c.

The increase of polarizability with atomic weight is mainly due to the increase in the number of electrons per atom. Also, relation 15.1 has been found to hold very accurately.

Agreement of equation 15.5 with experiment holds for many other non-polar substances over a very wide range of densities. Thus we notice with van Vleck [V3] that for O_2 the right-hand side of 15.5 has the magnitude 3·869 for the gas and an almost identical value 3·878 for the liquid despite the fact that the densities differ by a factor of over a thousand. Similar agreement is found in many other cases, and since the Maxwell relation holds, ϵ in 15.5 can be replaced by n^2. In nitrogen, for instance, it was found from measurements of the refractive index that the right-hand side of 15.5 varied by less than 1 per cent. in a range of pressures between 1 and 2,000 atmospheres.

In judging the significance of such a striking agreement it should be noted that in spite of the large variation in the magnitude of $\epsilon-1$, the largest values of this quantity are usually smaller than unity. In this case, even with the assumption of complete absence of interaction between molecules, there is fair agreement with experiment. This assumption (cf. 6.14, using 15.4 and 15.2) leads to

$$\epsilon-1 \simeq \frac{3d}{W}p. \tag{15.6}$$

In order to compare 15.5 and 15.6 we note that if we introduce the molecular volume v given by

$$v = \frac{W}{d}, \tag{15.7}$$

equation 15.5 is equivalent to

$$\epsilon-1 = \frac{3p/v}{1-p/v}. \tag{15.8}$$

Thus

$$\epsilon-1 = \frac{3p}{v}\left(1+\frac{p}{v}+\left(\frac{p}{v}\right)^2+...\right) \quad \text{if } \frac{p}{v} = \frac{\epsilon-1}{\epsilon+2} < 1. \tag{15.9}$$

The first term in this series is identical with 15.6. The higher terms which are characteristic of the Clausius–Mossotti formula are usually small. Thus in the above-mentioned case of nitrogen p/v is only slightly larger than 0·1 at 2,000 atmospheres. Compared with equation 15.6 the Clausius–Mossotti formula introduces a correction of only about 10 per cent.

Kirkwood [K3] has investigated the influence of the anisotropy of the polarizability of molecules and found that 15.9 has to be replaced by

$$\epsilon-1 = \frac{3p}{v}\left(1+(1+\sigma)\frac{p}{v}+...\right), \tag{15.10}$$

where σ is a quantity measuring the anisotropy of the polarizability. The effect of this and other corrections on the magnitude of ϵ are usually small, since they affect the small second-order term only.

There are a great number of materials which according to the classification of § 14 are polar, but which, for all practical purposes behave like non-polar substances. They comprise molecules which contain a number of polar groups in such a way that their resultant dipole moment vanishes. The infra-red polarizability of these molecules is very often so small that its contribution to the dielectric constant is negligible. Important examples are the paraffins, and in the following we shall show why the dipole moment of a paraffin molecule vanishes.

A molecule of a normal paraffin CH_3—$(CH_2)_n$—CH_3 consists of a chain of CH_2 groups with a CH_3 group at each end. The bond angle C—C—C is very nearly equal to the tetrahedral angle $\theta = 2\cos^{-1}(1/\sqrt{3}) \simeq 109°$. In solids the chain forms a plane zigzag (e.g. § 16, Fig. 28), but in liquids and gases this is not the case in general. Here rotation around the C—C bonds as axes may occur, but it will be assumed that this does not alter the bond angle. Each CH_2 and CH_3 group has a dipole moment whose direction will be discussed on the assumption that all four bonds of a C-atom make angles of $2\cos^{-1}(1/\sqrt{3})$ with each other.† Thus, if a C-atom is considered to occupy the centre of a regular tetrahedron, the bonds are directed towards the vertices. Now, according to § 14, the C—C bonds should have no dipole moment in contrast to C—H bonds which will have a moment. The resultant moment of a CH_2 group is then always in the C—C—C plane (cf. Fig. 20) bisecting the C—C—C angle, whereas the CH_3 moment has the direction of the adjoining C—C bond. To show this we consider a cube with a C-atom at its centre (Fig. 21). Then the four bond directions point towards four of its corners which are not neighbours. These corner points form the vertices of a regular tetrahedron. Thus, if μ is the C—H moment, the CH_2 moment is $2\mu\cos\theta/2 = 2\mu/\sqrt{3}$. This may be considered as composed of two dipoles of strength μ in the directions of the C—C bonds (dotted arrows in Fig. 20). Furthermore, the moment of a CH_3 end-group is also equivalent to a dipole μ in the direction of its adjoining C—C bond. For the resultant of

† It should be noted that in Fig. 28 the distance refers to CH_2 groups as found from X-ray measurements and not to the C—C distance.

any pair of H-atoms of the CH_3 group is a dipole of strength $2\mu/\sqrt{3}$ in the direction perpendicular to a surface of the cube, as has just been shown (this assumes that C—H dipoles in a CH_2 and in a CH_3 group are equal). The resultant thus forms an angle $\cos^{-1} 1/\sqrt{3}$ with the diagonal in which the residual dipole lies, and thus contributes $(1/\sqrt{3})(2\mu/\sqrt{3}) = 2\mu/3$ to the CH_3 dipole. Now three CH_2 groups can be selected from CH_3, but then each CH dipole is counted twice. Thus the moment of the

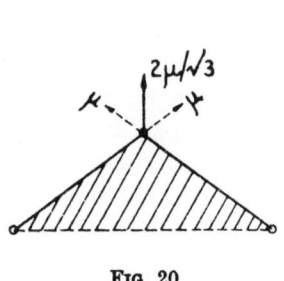
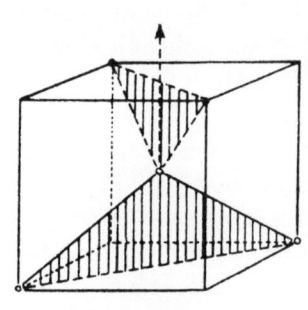

Fig. 20.   Fig 21.

Fig. 20. The full arrow represents the dipole of a CH_2 group seen to lie in the C—C—C plane of a paraffin molecule. It can be decomposed into the two dotted arrows in the C—C directions.

Fig. 21. A C-atom of a paraffin molecule with its four neighbours, two C- and two H-atoms. The former are indicated by rings, the latter by dots; the arrow indicates the dipole of the group. The lines joining the central atoms with its neighbours indicate the directions of the latter rather than their distance; actually the C—H distance is smaller than the C—C distance.

CH_3 group is $\frac{3}{2}(2\mu/3) = \mu$. It follows that the paraffin chain can be considered as composed of C—C rods with a dipole μ on each end. These end dipoles point in opposite directions, so that the dipole moment of each rod, and hence of the whole chain, vanishes exactly.

As an example we mention pentane C_5H_{12} which at 30° C. under a pressure of 1 atmosphere has a dielectric constant $\epsilon = 1\cdot 82$. The refractive index is $n = 1\cdot 36$, i.e. $n^2 = 1\cdot 85$ which shows that the Maxwell relation 15.1 is fulfilled—probably within experimental error. Hence the contribution of the infra-red polarization must be very small. The following table of measurements by Danforth [D1] shows that the Clausius–Mossotti formula also holds to a good approximation.

Pressure	ϵ	$(\epsilon-1)/(\epsilon+2)$	$1/v$	$p = v(\epsilon-1)/(\epsilon+2)$
1 atm.	1·82	0·216	0·613	0·356
12000 atm.	2·33	0·308	0·907	0·339

Again it can be seen that (cf. 15.9) $p/v \ll 1$ so that the zero approximation of 15.9 accounts for more than two-thirds of $\epsilon-1$. More accurate comparison with equation 15.10 shows that a value $\sigma \simeq -0\cdot05$ is required to account for the experimental data. This shows that the next approximation—i.e. the first one which is significant for the Clausius–Mossotti formula—deviates by only 5 per cent. from 15.9.

For an example exhibiting a larger contribution of infra-red polarization we turn to the measurements of Michels and Hamers [M2] and Michels and Kleerekoper [M3] on CO_2. This molecule is rectilinear and has, therefore, no dipole moment. Dielectric measurements give the value 7·5 for $v(\epsilon-1)/(\epsilon+2)$; from refractive index measurements, on the other hand, the value 6·7 is found for $v(n^2-1)/(n^2+2)$. Hence about 10 per cent. of the static polarization is due to infra-red contributions in this case.

Relatively large infra-red contributions are obtained in dipolar solids at low temperatures at which dipoles are frozen in, so that these substances behave in a similar manner to non-polar materials (temperature-independent dielectric constant in a certain range). As an example, the case of ketones will be discussed in more detail in § 17.

16. Dipolar substances

Gases and dilute solutions

Provided the pressure is not too high, the static dielectric constant ϵ_s of a dipolar gas satisfies the relation $\epsilon_s - 1 \ll 1$. Therefore ϵ_s should satisfy equation 6.15, which, with the help of 15.2 can also be written as

$$\epsilon_s - \epsilon_\infty = \frac{4\pi}{3} 6\cdot02 \times 10^{23} \frac{d}{W} \frac{\mu_v^2}{kT}, \qquad (16.1)$$

or, inserting numerical values and expressing μ_v in Debye units (cf. § 6) and T in degree Kelvin,

$$\epsilon_s - \epsilon_\infty = 1\cdot 83 \times 10^4 \frac{d}{W} \frac{\mu_v^2}{T}. \tag{16.2}$$

This equation, according to § 6, has been derived on the assumption that the interaction between dipoles is negligible. This

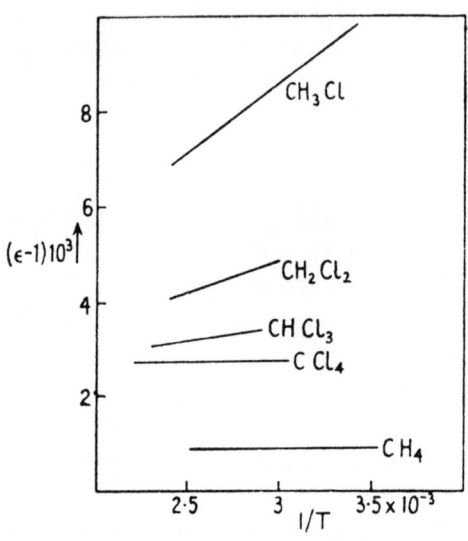

Fig. 22. Temperature-dependence of the dielectric constant of various gases according to Sänger [S1].

involves an error of the order of $(\epsilon_s-1)^2$, as can be seen by comparison with formulae where the interaction has not been neglected (e.g. 6.36).

Since no assumptions about the interaction have to be made for the derivation of 16.1 or 16.2, measurement of the temperature-dependence of the static dielectric constant of gases should make possible a reliable determination of the dipole moment μ_v of free molecules. As an example Fig. 22 shows the results of measurements of the dielectric constants of CH_4, CH_3Cl, CH_2Cl_2, $CHCl_3$, and CCl_4 gas plotted against the inverse temperature $1/T$ (from Sänger, S1). Since the highest value of $\epsilon_s - 1$ is only about 10^{-2}, the above equations should hold very well. As required, the experimental points lie on a straight line whose slope allows us to calculate the dipole moment with the help of 16.1 or 16.2 if the density is also measured.

The results are also of interest in the field of stereochemistry. Thus the fact that the dipole moments of methane (CH_4) and of carbon tetrachloride (CCl_4) both vanish (because ϵ_s is independent of T) is evidence in favour of a molecular structure in which the four hydrogen (or chlorine) atoms form the corners of a regular tetrahedron with the carbon atom at the centre.

Figure 22 also permits a determination of ϵ_∞, which is obtained by extrapolating the straight lines to $1/T = 0$. The value obtained should be slightly larger than the square of the optical refractive index n. But, as already mentioned in § 15, the difference $\epsilon_\infty - n^2$ is very small for most molecules; in gases it is probably too small to be measurable using present experimental techniques.

There should be no appreciable frequency dependence of the dielectric constant, and hence no loss, up to the absorption frequencies of the molecules, which often lie in the infra-red region. For a number of molecules absorption begins at the still longer wave-lengths of the ultra-short electric wave region. Such gases will then show resonance absorption in this region following the laws derived in § 13. Here we merely wish to discuss briefly the shape of the absorption spectrum of ammonia which, following Cleeton and Williams [$C2$], has been investigated in the centimetre region in great detail by Bleaney and Penrose [$B3$]. Ammonia, NH_3, forms a pyramidal molecule and the absorption is connected with the swinging of the nitrogen atom through the plane of the three hydrogen atoms. The spectrum shows considerable structure, which is resolved at pressures below 5 cm. Hg, as shown in Fig. 23. The shape of a single absorption line should be given by equation 13.11, provided the concept of a single relaxation time τ can be applied. Assuming $1/\tau$ to be proportional to the gas pressure, the shape of the absorption spectrum at a pressure of 10 cm. of mercury has been calculated by Bleaney and Penrose with the help of 13.11 using data from measurements at 0·5 mm. Figure 24 shows excellent agreement with the measured absorption. At a pressure of 60 cm., however, agreement is no longer so good.

We shall now turn to dilute solutions of dipolar molecules in non-polar substances. The concentration will be assumed to be

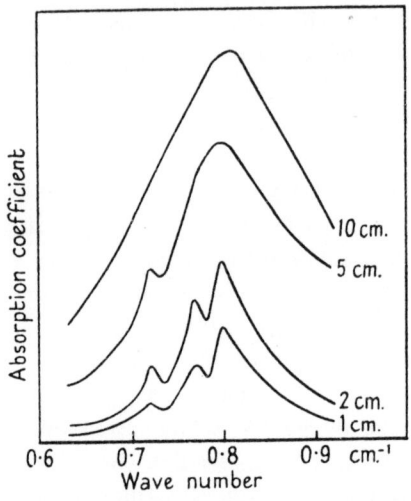

Fig. 23. The ammonia absorption spectrum at 1, 2, 5, and 10 cm. pressure according to Bleaney and Penrose [B3].

sufficiently small so that interaction between dipoles may be neglected. It then follows from equation 6.19 that the static dielectric constant ϵ_s satisfies an equation similar to those of gases, but with an effective dipole moment which depends on the nature of the solvent and on the structure of the dipolar molecule. This effective dipole moment can be calculated in a simple way only when the molecule can be approximated by a point dipole at the centre of a sphere with refractive index n. In this case equation 6.24 should hold. In general, we may put ($\epsilon_0 =$ dielectric constant of solvent; $N_0 =$ number of dipole molecules per unit volume)

$$\epsilon_s - \epsilon_\infty = \frac{4\pi\mu_v^2 N_0}{3kT}\left(\frac{\epsilon_0+2}{3}\right)^2 (1-\gamma)^2, \qquad (16.3)$$

where γ is unknown except in the case of spherical molecules for which, according to 6.24,

$$\gamma = \frac{2(\epsilon_0-1)(\epsilon_0-n^2)}{(2\epsilon_0+n^2)(\epsilon_0+2)} \quad \text{for spherical molecules.} \quad (16.4)$$

It follows that $|\gamma|$ is usually much smaller than unity, but, in view of the factor ϵ_0-n^2, may be either positive or negative.

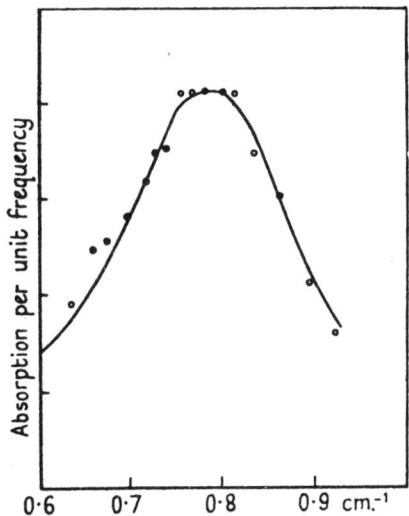

Fig. 24. Calculated and measured absorption spectrum of ammonia at 10 cm. pressure according to Bleaney and Penrose [B3].

The main conclusion to be drawn is that by using experimental values for the temperature-dependence of ϵ_s it is not possible to calculate the dipole moment μ_v of the free molecule but only the quantity $\mu_v(1-\gamma)$ which differs from μ_v, though usually only by a small amount.

Besides the solvent effect, one of the main differences between gases and dilute solutions is their behaviour in alternating fields. The interaction between the dipolar molecules and the molecules of the solvent leads to an energy loss whose maximum at most temperatures falls within the frequency region of electric waves. This affords a new way of obtaining the important quantity $\epsilon_s-\epsilon_\infty$ without measuring the static dielectric constant

ϵ_s; for if the dielectric loss, and hence the imaginary part $\epsilon_2(\omega)$, of the complex dielectric constant $\epsilon(\omega)$ (cf. § 2) is known as a function of frequency, $\epsilon_s - \epsilon_\infty$ is obtained from equation 2.18, which may also be written as

$$\epsilon_s - \epsilon_\infty = \frac{2}{\pi} \int \epsilon_2(\omega) d(\log \omega), \qquad (16.5)$$

and which holds independently of the special shape of the function $\epsilon_2(\omega)$. The integration has to be carried out over the whole range of frequencies in which $\epsilon_2(\omega)$ has appreciable values. The present method can be expected to yield more accurate values for $\epsilon_s - \epsilon_\infty$ than measurement of the static dielectric constant because ϵ_∞ is nearly equal to the dielectric constant ϵ_0 of the solvent and $\epsilon_s - \epsilon_\infty$ is small compared with ϵ_0 in the case of dilute solutions. Very high accuracy in the measurement of ϵ_s would therefore be required to give reasonably accurate values of $\epsilon_s - \epsilon_\infty$.

The investigation of dilute solutions permits a study of the behaviour of single molecules independent of dipolar interaction; and the results obtained will be of interest in attempts to understand the properties of less dilute solutions, or of pure dipolar substances. To obtain full information the dielectric loss should be measured over a large range of frequencies at different temperatures. Absorption can usually be described by the Debye formulae (§ 10) or by a generalization (§ 12). Frequency measurements will establish whether the Debye equations hold; if they do, they will lead to a value for the relaxation time, but otherwise they will permit an estimate of the width of a band of relaxation times. Temperature measurements will then yield the temperature-dependence of the relaxation time—or of the width and position of the band of relaxation times. No complete system of measurements is available at present.

As an example, we shall now discuss the experiments by Jackson and Powles [J3] who have measured at one temperature (19° C.) the frequency dependence of dielectric loss of dilute solutions of dipolar molecules in benzene and in paraffin. In Fig. 25 the experimental values of $\tan \phi$ (ϕ = loss angle) for benzophenone solutions (1 gm./100 c.c. at 19° C.) are plotted against frequency. For the solution in benzene these values lie on a curve which represents the Debye formula. According to 10.17, 10.22, and 10.23, making use of 10.29, this formula can be written in terms of the angular frequency ω_m at which the maximum loss-angle ϕ_m occurs:

$$\tan\phi = \tan\phi_m \frac{2\omega/\omega_m}{1+(\omega/\omega_m)^2}, \qquad (16.6)$$

where
$$\tan\phi_m = \frac{1}{2}\frac{\epsilon_s-\epsilon_\infty}{\epsilon_s}, \qquad \epsilon_s-\epsilon_\infty \ll 1. \qquad (16.7)$$

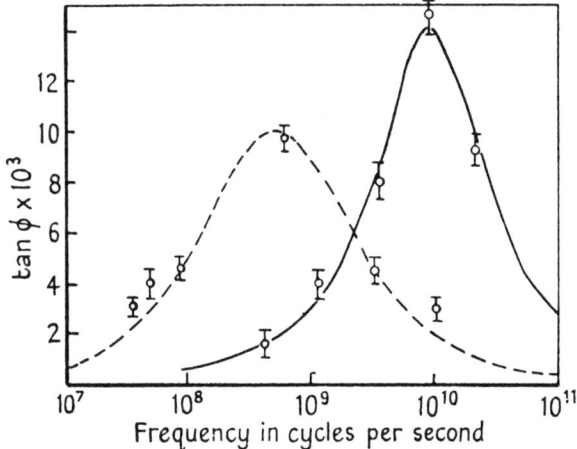

FIG. 25. Loss angle of dilute solutions of benzophenone in benzene (maximum near 10^{10} cycles) and paraffin; experimental values according to Jackson and Powles [J3]. The full line represents a Debye curve (equation 16.6, or 16.8 with $\beta = 1$); the dotted line represents equation 16.8 with $\beta = 5$.

From 10.22 the relaxation time is then given by $\tau = 1/\omega_m$.

In contrast to benzene the solution of benzophenone in paraffin leads to a broader absorption curve than could be satisfied by a Debye curve. According to § 12 this can be described by introducing a whole range of relaxation times. If, in particular, we introduce a model where the relaxation times are distributed between the two values τ_0 and τ_1 according to the distribution function 12.13, then $\tan\phi$ can be obtained after division by $\epsilon_1(\omega) \simeq \epsilon_s$ (cf. 10.29) from 12.16 or from 12.18 and 12.19. Thus with the use of 12.17,

$$\tan\phi = \tan\phi_m \frac{\tan^{-1}(\omega\beta/\omega_m) - \tan^{-1}(\omega/\beta\omega_m)}{\tan^{-1}\beta - \tan^{-1}(1/\beta)}, \quad (16.8)$$

where
$$\tan\phi_m = \frac{1}{2}\frac{\epsilon_s - \epsilon_\infty}{\epsilon_s}\frac{\tan^{-1}\beta - \tan^{-1}(1/\beta)}{\log\beta}, \quad (16.9)$$

and
$$\beta = \sqrt{(\tau_1/\tau_0)} \geqslant 1. \quad (16.10)$$

For $\beta = 1$, equations 16.8 and 16.9 are identical with the Debye equations 16.6 and 16.7. For $\beta > 1$, however, the curves are broader and flatter. Figure 25 shows that the curve with $\beta = 5$ satisfies the experimental points fairly well; appreciable deviations occur only at frequencies very far from ω_m where $\tan\phi$ is relatively small. This is rather to be expected, as follows from the discussion in § 12 following equation 12.19, because the distribution function 12.13 can be considered to represent position and width of the distribution of relaxation times but not finer details.

It will be of interest to see how the time $1/\omega_m$ which in the Debye case represents the relaxation time is now connected with the two limits τ_0 and τ_1, between which the relaxation times are distributed. According to 12.17,

$$1/\omega_m = \sqrt{(\tau_0\tau_1)} = \beta\tau_0 = \tau_1/\beta. \quad (16.11)$$

Thus, in the present case $1/\omega_m = 5\tau_0 = \tau_1/5$.

Since no experiments at other temperatures are available at present, it is not yet possible to make use of the temperature-dependence of the relaxation times to discuss the relaxation mechanism. On the assumption of a model in which the relaxation time is determined by the frequency of jumps of a dipole over potential barriers (cf. § 11), the factor β allows us to calculate the maximum difference v_0 in height of these potential barriers on the assumption that they are equally distributed between the values H_0 and $H_0 + v_0$ (cf. 12.7).

Thus from 12.17, using 16.10,

$$v_0/kT = \log(\tau_1/\tau_0) = 2\log\beta. \quad (16.12)$$

In our case, therefore, $v_0/kT = \log 25 \simeq 3$.

It would be of interest to investigate the variation with temperature of the ratio τ_1/τ_0 as well as that of τ_1 (or τ_0) itself. Measurements at various temperatures were carried out by Whiffen and Thompson [W2], though unfortunately, up to the time of writing, the range of frequencies has been rather restricted. The dielectric loss in solutions of chloroform in heptane seems to follow a Debye curve as indicated in Fig. 26, but it

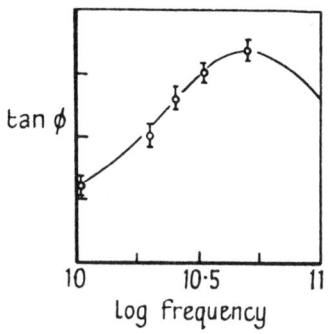

FIG. 26. Loss angle of a dilute solution of chloroform in heptane according to Whiffen and Thompson [W2]; the curve represents the Debye formula 16.6.

must be realized that in the available frequency range there is only a variation of a factor of three in $\tan\phi$. The temperature was varied in a range between $-70°$ C. and $80°$ C. It was found that the logarithm of the relaxation times is proportional to $1/T$. A similar result was obtained for solutes, as shown in Fig. 27. To test the validity of the Debye model (§ 11), according to which the temperature variation of τ should be the same as that of the viscosity η of the solvent, $\log\eta$ is also plotted in the same figure. Putting
$$\tau \propto e^{H_\tau/kT}, \qquad \eta \propto e^{H_\eta/kT}, \qquad (16.13)$$
Debye's model according to 11.32 requires $H_\tau = H_\eta$. The experimental values are as follows:

	H_τ in k.cal.
α-Bromo-naphthalene	1·8
Methyl benzoate	1·8
Camphor	1·7
Chloroform	1·5

On the other hand, for heptane $H_\eta = 2\cdot 0$, which thus is larger than H_r for all the above molecules. This indicates that in the above cases the Debye model probably does not apply exactly.

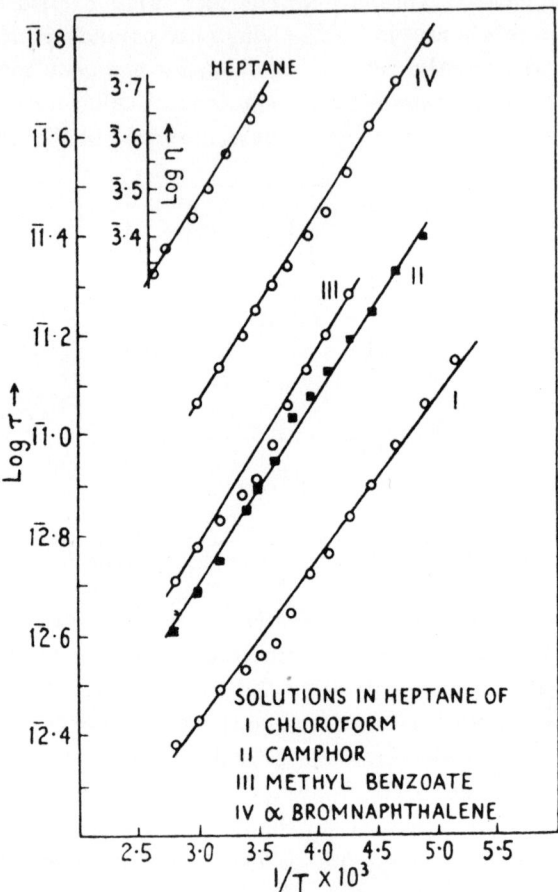

Fig. 27. Temperature-dependence of the relaxation time τ of various solutions in heptane, and of the viscosity η of heptane, according to Whiffen and Thompson [$W2$].

In all experiments on dilute solutions the value of $\epsilon_s - \epsilon_\infty$ can be obtained from dielectric loss measurements by carrying out the integration indicated in equation 16.5. It should be remembered that for dilute solutions both quantities $\epsilon_1(\omega)$, ϵ_s are nearly equal to ϵ_0, the dielectric constant of the solvent, so that (cf.

2.5) $\epsilon_2(\omega)$ can be replaced by $\epsilon_0 \tan\phi$. From $\epsilon_s - \epsilon_\infty$, then, the dipole moment μ_r can be obtained according to 16.3 if γ can be neglected. This method was first used by Sillars [*S6*] in the case of dilute solutions of dipolar long-chain molecules in solid paraffin-wax. Although equation 16.4 for γ is certainly not applicable in this case, it nevertheless indicates that γ should be particularly small if the square n^2 of the refractive index of the pure solute is nearly equal to the dielectric constant of the solvent. For long-chain molecules this should be very well fulfilled.

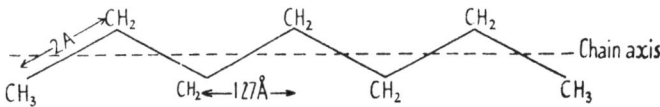

Fig. 28. Paraffin chain. The indicated distances refer to X-ray scattering centres of CH_2 groups according to Muller [*M6*]; they are slightly different from C—C distances.

The investigation of the solid solutions mentioned above are of great interest for the elucidation of the mechanism of dielectric loss and will, therefore, be discussed in greater detail. We shall start with a discussion of the structure of substances built up of long-chain molecules. Most of these structures can be derived from that of paraffins, which was investigated by A. Muller [*M6*]. A paraffin molecule, in a crystal, consists of a plane zigzag with a CH_2 group at each corner and a CH_3 group at each end (cf. Fig. 28). The distance between neighbouring CH_2 groups is about 2 A (1 A = 10^{-8} cm.), while the distance of their projections to the chain axis is about 1·25 A. In a paraffin crystal the chains are arranged in layers whose thickness is approximately equal to the chain length. Within such a layer the molecules form rectangular cells with side length a, b, c, where $a \simeq 5$ A, $b \simeq 7\cdot5$ A, and c is slightly larger than the chain axis and is parallel to it. Figure 29 shows how the chain planes cross the a–b plane which is perpendicular to the chain axes. It is of importance to notice that in the subsequent layer the whole arrangement is shifted by about 1 A in the direction of the b-axis. This is indicated in Fig. 29 by the dotted lines.

A paraffin molecule carries no dipole moment, as has already been mentioned in § 15. Now suppose that some dipolar long-chain molecules, e.g. esters or ketones, are dissolved in a paraffin crystal. This means that some of the paraffin molecules are replaced by dipolar long-chain molecules. These latter can be usually derived from paraffin molecules. A ketone molecule, for instance, is obtained from a paraffin molecule by replacing one CH_2 group by a CO group. This latter carries a dipole moment whose direction is perpendicular to the chain axis and probably

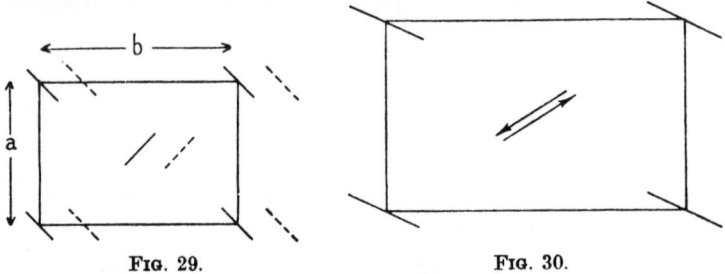

Fig. 29. Section of paraffin crystal perpendicular to chain axis indicating the chain planes, according to Muller [M6]. The dotted lines indicate the section of chain planes in a subsequent layer.

Fig. 30. The two equilibrium positions of a ketone molecule—indicating its dipole directions—replacing a paraffin molecule of longer chain length. The size of the rectangular cell is the same as in Fig. 29.

lies in the chain plane. Also assume that the length of the ketone molecule is shorter than that of a paraffin molecule so that it can easily replace a paraffin molecule. There should then be two possible directions for the dipole which differ by 180°, as indicated in Fig. 30. For, if the ketone molecule is shorter than the paraffin molecule, the former should still fit into the structure if it is turned through 180° and shifted by the length of one link along the chain axis (cf. Fig. 31). In a dilute solid solution such a dipolar molecule has thus two equilibrium positions with opposite dipole directions. From the similarity of the two positions it can be assumed that the energy of the molecule is the same in both.

In general, a molecule will oscillate about either of the two equilibrium positions. Occasionally, due to a thermal fluctuation, it will acquire sufficient energy to turn into the other equilibrium position. Such a substance thus may well be described by the high-temperature model of dipolar solids discussed in § 11, where a dipole was supposed to have two equilibrium positions separated by a potential barrier; the minimum energy required to lift it over the barrier was denoted by H. The energy loss of such a substance is described by the Debye formulae (e.g. 16.6 and § 10) and the relaxation time is given by equation 11.3 (cf. also 11.8).

Loss measurements on dilute solutions of the type described above have been carried out by W. Jackson [$J1, 2$], Sillars [$S6$],

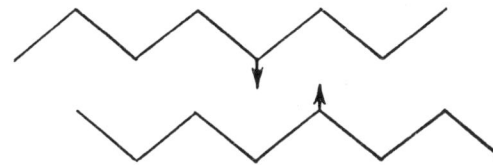

Fig. 31. Result of turning a ketone molecule by 180° and shifting it by one chain length. The arrow indicates the dipolar CO group.

and Pelmore [$P2$]. These authors find that the dielectric loss curves satisfy the Debye equations fairly well, though not exactly. This slight deviation from the Debye equations may be due to the fact that a dipolar molecule which is shorter than a paraffin molecule, say by z links, has z possible locations (apart from the two dipole directions) corresponding to different positions on the c-axis. Its energy in these positions may not be exactly the same, and the same may be true of the energy H required to lift a molecule over the potential barrier. This would lead to a distribution of relaxation times. It should be realized that in calculating the relaxation time τ according to the Debye formula 10.22 some error is introduced (cf. 16.11), though it may be negligible if the variation of $\log \tau$ with temperature, and not the absolute value of τ, is considered.

The most interesting result of these experiments is the measurement of the dependence of the relaxation time on the chain length of the dipolar molecule. If the molecule is considered to be rigid, and if therefore each link has to be lifted over the barrier at the same time, the total energy H_m required to lift a whole molecule of chain length m over the barrier should increase in proportion to m. Thus, since according to 11.3

$$\log \tau = \text{constant} + H_m/T, \qquad (16.14)$$

the experimental values of $\log \tau$ plotted against m should lie on a straight line. This is not the case. Actually molecules are not rigid but have a certain flexibility, and, if this is taken into account (cf. reference $F5$), satisfactory agreement with experiment is obtained. The influence of the flexibility on the value of H_m can easily be discussed qualitatively. Suppose for a moment that the chain is rigid. Then the whole chain has to be lifted over the potential barrier at the same time. Thus if H_1 is the energy required to lift a single link over the barrier, the total energy H_m would be equal to mH_1 where m is the number of carbon atoms in the chain. Actually, however, the chain has a certain flexibility. This allows the chain to be gradually lifted over the potential barrier so that a total energy less than mH_1 is required. On the other hand, a twisting of the molecule requires a certain energy; but this energy decreases with increasing chain length for a given angle of twist of the two ends of the chain. Thus for short chains $H_m \propto m$, but for long chains H_m will tend to a constant value independent of the chain length. A simple calculation [$F5$] yields†

$$H_m = H_1 m_0 \tanh \frac{m}{\bar{m}}, \qquad (16.15)$$

where $\bar{m}$ is a constant which separates short chains ($m < \bar{m}$, little torsion) from long chains ($m > \bar{m}$, appreciable torsion). Comparison with experiments shows that $H_1 \simeq 1/30$ e-volt, while $\bar{m} = 26$, both reasonable values. To obtain τ a further constant is required according to equation 16.14. Thus three experimentally determined constants are included in the final formula (τ in seconds)

$$\log 2\pi\tau = -50{\cdot}4 + \frac{13800}{T} \tanh \frac{m}{26}. \qquad (16.16)$$

Here the relaxation time τ depends on two parameters, the temperature T and the chain-length m. The formula holds, however,

† Note that $\tan hx = (e^x - e^{-x})/(e^x + e^{-x})$.

only if m is smaller than the chain length of the paraffin in which the dipolar chains are dissolved. Figure 32 shows that the results obtained from equation 16.16 agree well with the experimental results. The expression $\tanh(m/26)$ is plotted as a function of $m/26$ and is compared with the experimental values of $(\log 2\pi\tau + 50 \cdot 4)T/13800$ for four different chain-lengths ($m = 20$, 22, 24, 32). Two of the constants in equation 16.16 have been determined from the experimental values for $m = 20$ and

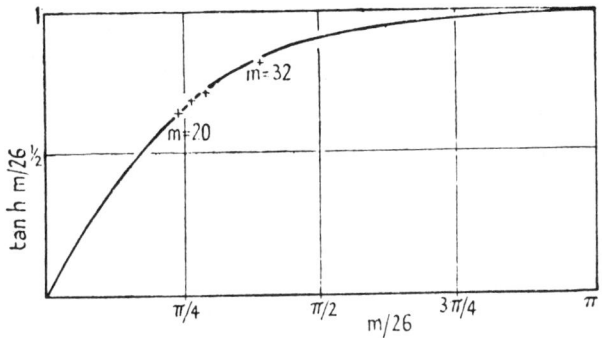

Fig. 32. Comparison of the theoretical dependence of the relaxation time on chain length with experimental values, according to [$F5$].

$m = 22$, while the third was determined from the temperature-dependence of $m = 20$. For $m = 24$ and 32 no more adjustable constants are available, which shows that formula 16.16 leads to a correct dependence of the relaxation time on chain lengths.

From the numerical value $-50 \cdot 4$ for the constant in equation 16.14 (comparing with 16.16) it might appear that the order of magnitude of the constant A of equation 11.3 (cf. also 11.8) could be estimated. For assuming for the frequency of oscillation $\omega_a/2\pi \sim 10^{12}$ one finds $\log A \sim -25$. Not much significance should be attributed to this value, however, in view of the uncertainties connected with the experimental determination of the absolute value of τ; for, as mentioned above, the deviation of the loss curve from a true Debye curve brings with it some uncertainties in determining τ which cannot be adjusted by applying equation 10.22 as in the Debye case. This would probably influence the value of A to a much greater extent than the less sensitive H_m.

In all examples discussed above the concentration of dipolar molecules was assumed to be so small that dipolar interaction would be neglected. In this range, therefore, the quantity $\epsilon_s-\epsilon_\infty$ as well as dielectric loss should increase proportionally with the concentration. For further increases of concentration the effects of dipolar interaction should start to appear, and an investigation of the deviations of $\epsilon_s-\epsilon_\infty$ from equation 16.3 would be of great theoretical interest. In particular, it would be of interest to see to what extent and in which range of concentration the Onsager formula 6.38 can account for such deviations. This would be of particular interest for a solute which in the pure state does not satisfy the Onsager formula for the temperature range contemplated.† It will be remembered (§ 8) that the Onsager formula takes into account the long-range dipolar forces, but neglects short-range forces. Then with increasing concentration we should expect three stages: (i) all interaction can be neglected, leading to equation 16.3 for $(\epsilon_s-\epsilon_\infty)$; (ii) long-range dipolar forces have to be considered, but short-range forces can still be neglected, leading to the Onsager formula 6.38; (iii) short-range forces have to be considered as well. This leads to the Kirkwood formula 8.14.

Of very great importance would be the investigation of the dependence on concentration of the shape of the dielectric loss-frequency curve. The deviations from a Debye curve would probably increase with increasing concentration owing to the interaction between dipoles. As yet no theoretical investigations on this subject have been carried out.

17. Dipolar solids and liquids

Survey

From the developments of §§ 7 and 8 it follows that no formula exists which represents in a simple manner the dielectric

† Measurements on the dependence of ϵ_s on concentration have been carried out, of course. They do not extend, however, to the region of very low concentration where $\epsilon_s-\epsilon_\infty$ is best obtained from loss measurements with the help of 16.5.

behaviour of dipolar solids and liquids. Qualitatively, however, most of these substances behave in a very similar way. At very low temperatures in the solid state all dipoles are frozen in, and if the solid has not become permanently polarized the dipoles will not contribute to the dielectric constant. In this region ($T \to 0$), therefore, the dielectric constant $\epsilon_s(0)$ should be nearly independent of temperature, and its magnitude should be approximately equal to that of the dielectric constant $\epsilon_\infty(T)$ at a higher temperature but at a frequency which is so high that the dipoles do not take part in the polarization. This equivalence holds only, of course, if the volume of the substance is the same at both temperatures, and it includes the assumption that upon turning a molecule from one equilibrium position into another its contribution to the high-frequency dielectric constant is not altered. Then

$$\epsilon_s(0) \simeq \epsilon_\infty(T) = n^2 + \Delta\epsilon, \quad \text{constant volume.} \qquad (17.1)$$

In this formula it has been indicated that ϵ_∞ is composed of the square of the optical refractive index n (cf. § 15), and of a term $\Delta\epsilon$ due to elastic displacement of nuclei or of dipoles. The corresponding polarization was called infra-red polarization in § 14. As pointed out in § 15, this type of polarization is frequently negligibly small in non-dipolar substances. In dipolar substances $\Delta\epsilon$ usually has small but noticeable values, mainly due to a displacement of the equilibrium directions of the dipoles by the field. The frequency of the rotationary oscillations which the dipoles carry out about their equilibrium positions is normally in the far infra-red region, but there are indications that in large molecules it may approach the centimetre region (cf. reference F9). Such substances would thus show resonance absorption (§ 13) at ultra-high frequencies. Substances which are expected to show such absorption have been discussed by Szigeti [S11] and a further theoretical investigation on the expected frequency-dependence of the absorption has been carried out by Huby [H3] leading to a generalization of equation 13.11. In this connexion it should be mentioned that Girard and Abadie [G3] found that normal liquid long-chain alcohols absorb in the centimetre region besides showing Debye absorption at longer wave-lengths. It was suggested by Magat [*M1*] that this is an example of resonance absorption. Further experiments would be required to establish the validity of this interpretation.

Let us now turn to the regions of higher temperatures and lower frequencies where dipole orientation plays an important role. Within creasing temperature an increasing number of

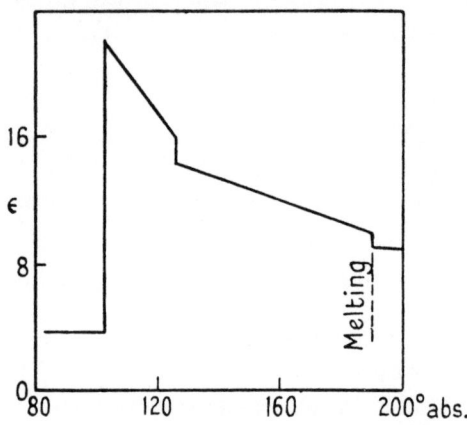

FIG. 33. Temperature-dependence of the dielectric constant of solid hydrogen sulphide according to Smyth and Hitchcock [*S9*] showing an order-disorder transition near 103° abs. A further transition and melting are of little influence.

dipoles can turn into other equilibrium directions, as pointed out in § 8. The static dielectric constant ϵ_s will thus increase with temperature, at first according to the relation

$$\epsilon_s - \epsilon_\infty \propto e^{-V(0)/kT}, \tag{17.2}$$

as follows from 8.17 ($V(0)$ is a constant energy discussed in § 8). At higher temperatures the rate of rise of ϵ_s gets increasingly larger until at the temperature T_c the solid carries out an order-disorder transition beyond which ($T > T_c$) the dielectric constant ϵ_s falls with increasing temperature. The order-disorder transition may occur in the solid state; a number of examples are shown in Figs. 33 and 34. In this case there is usually no appreciable change of ϵ_s at melting. In other substances melting occurs before the order-disorder transition is completed. In these cases the decrease of ϵ_s starts at the melting-point.

As shown in the discussion of entropy in § 3, a positive value of $\partial \epsilon_s/\partial T$ indicates that the degree of order is decreased when the field is applied, while a negative value of $\partial \epsilon_s/\partial T$ indicates the opposite, namely, that the degree of order is increased by the application of the field. It follows that in a substance in which the dipole directions are completely disordered $\partial \epsilon_s/\partial T$ must be negative, whereas if the dipole directions are completely ordered

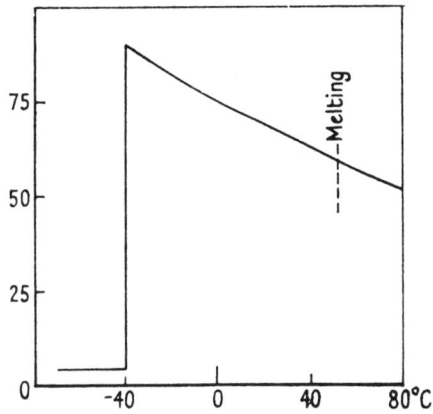

FIG. 34. Temperature-dependence of the dielectric constant of solid ethylene cyanide according to White and Morgan [$W4$] showing an order-disorder transition near $-40°$ C. without a discontinuity at melting.

$\partial \epsilon_s/\partial T$ must be positive. Thus from purely thermodynamical reasoning it follows that the transition from positive to negative values of $\partial \epsilon_s/\partial T$ must be connected with a change in the order of dipole directions. A large number of examples of such transitions has been reviewed in a recent article by Smyth [$S7$].

Dielectric loss due to dipole orientation is expected to occur at $T > T_c$ as well as for $T < T_c$ (Figs. 35 a and b give examples for the latter). Except very close to T_c the latter will, of course, be much smaller than the former because of the much smaller value of $\epsilon_s - \epsilon_\infty$. Thus (cf. Fig. 36) at any temperature we expect resonance loss in the far infra-red or ultra-high frequency region, and loss of the Debye type (but requiring a distribution of relaxation times) at longer wave-lengths. For the latter, no satisfactory theoretical treatment of the shape of dielectric

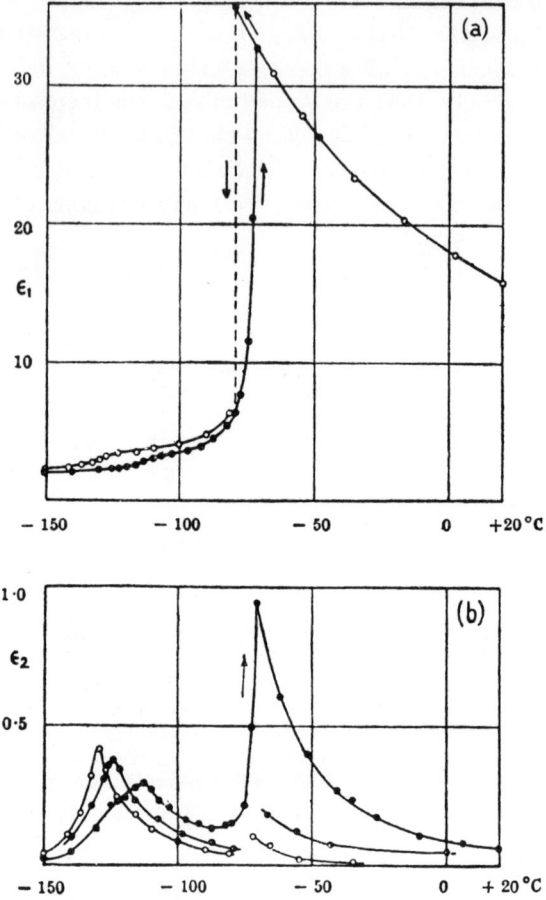

FIG. 35. (a) Dielectric constant ϵ_1, and (b) dielectric loss $\propto \epsilon_2$ of di-isopropyl ketone according to Schallamach [S4]; —O— at a frequency of 1·2 Mc./s., —◐— at 4·4 Mc./s., —●— at 20 Mc./s. Melting occurs near −73° C. and there is no previous transition.

loss-frequency curves exists at present. Compared with dilute solutions the difficulty consists in accounting for the influence

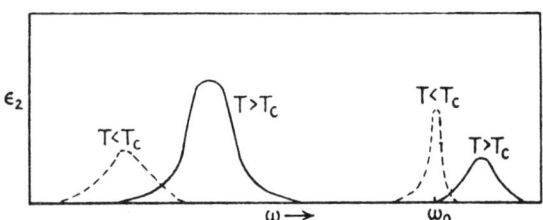

FIG. 36. Frequency-dependence of dielectric loss, ———— above and - - - - below the transition temperature T_c (schematically). $\omega_0/2\pi$ represents the lowest resonance frequency which is usually in the far infra-red or in the ultra-high electric frequency region.

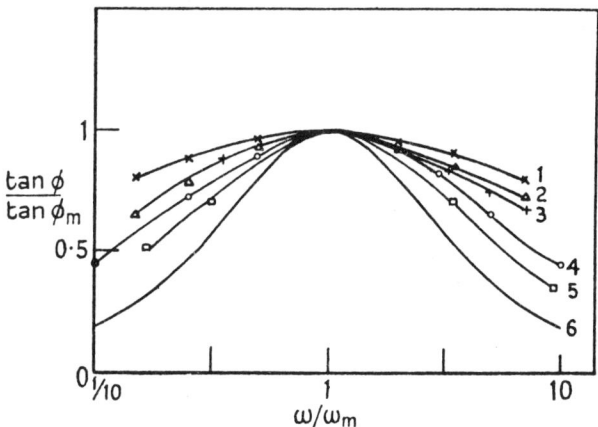

FIG. 37. Frequency-dependence of loss angle according to Hartshorn, Megson, and Rushton [*H1*] for various materials.
(1) Phenolic resins. (2) Benzyl alcohol resins. (3) Rubber-sulphur compounds (Scott, McPherson, and Curtis, *S5*). (4) Cetyl palmitate in paraffin wax (Sillars, *S7*). (5) Chlorinated diphenyl (Jackson, *J2*). (6) Represents a Debye curve; comparison with Fig. 14 shows that curves (1)–(5) can be represented by equation 12.19 with suitable values of the parameter τ_1/τ_0.

of interaction. This, one should expect, would lead to a broadening of the Debye curves similar to that obtained through the existence of a whole range of relaxation times (cf. examples in Fig. 37). The width of the latter ought to be larger in amorphous substances than in crystalline solids. It would be of interest, therefore, to investigate substances which at the same temperature exist in various states of order; e.g. amorphous substances which have been quenched from different temperatures, or substances existing at the same temperature as super-cooled liquids and as crystalline solids.

Analysis of $\epsilon_s \div T$ curves

Without any knowledge of the structure of a dielectric, interesting conclusions about the behaviour of its dipoles can be drawn from the experimental $\epsilon_s \div T$ curves with the help of our general theory. For from equation 7.39 it follows that we can obtain the quantity

$$\frac{4\pi}{3} \frac{N_0}{k} \overline{\mathbf{mm^*}} = (\epsilon_s - n^2) \frac{2\epsilon_s + n^2}{3\epsilon_s} T \qquad (17.3)$$

provided that the optical refractive index n is known in the correct region (cf. § 15). If this is not the case, we can make use of the fact that the infra-red contribution $\Delta\epsilon = \epsilon_\infty - n^2$ is usually small (except in ionic crystals) compared with n^2 and will replace n^2 by ϵ_∞ and hence, according to 17.1, by $\epsilon_s(0)$. Thus, if

$$B(T) = \frac{4\pi}{3} \frac{N_0}{k} \overline{\mathbf{mm^*}}, \qquad (17.4)$$

we find approximately

$$B(T) \simeq \{\epsilon_s(T) - \epsilon_s(0)\} \frac{2\epsilon_s(T) + \epsilon_s(0)}{3\epsilon_s(T)} T. \qquad (17.5)$$

This value 17.5 for $B(T)$ differs from the correct one by

$$T\Delta\epsilon \frac{\epsilon_s(T) + 2\epsilon_s(0) - \Delta\epsilon}{3\epsilon_s}, \qquad (17.6)$$

which is of the order $T\Delta\epsilon$. Using approximation 17.5, it is thus possible to obtain $B(T)$ from measurement of the static dielectric constant ϵ_s at various temperatures. Theoretically $B(T)$, according to 17.4, is proportional to $\overline{\mathbf{mm^*}}$, and the value given by 17.5 is thus constant (subject to the small correction 17.6) independent of temperature, if Onsager's formula holds, i.e. if the dipole moment of a molecule can be considered as a constant and if there are no short-range forces leading to mutual orientation of dipoles. Since for $T \to \infty$ all interaction can be neglected and the dipole orientation is completely random, a function $B(T)$ which increases with decreasing temperature points to an increasing tendency towards parallel orientation of dipoles. Similarly if $B(T)$ decreases with decreasing temperature the dipoles tend towards anti-parallel orientation. It should be pointed out again (cf. §§ 7 and 8) that this orientation is due to non-dipolar short-range forces. Long-range dipolar forces too lead to a certain tendency for dipole orientation. But averaging over all angles between dipole direction and radius vector, this orientation cancels in an isotropic substance. Since $\mathbf{m}^*$ is the average moment of a sphere if one of its units (molecules) has the moment $\mathbf{m}$, dipolar interaction alone leads to $\mathbf{m}^* = \mathbf{m}$, i.e. to a value of $\overline{\mathbf{m}\mathbf{m}^*}$ which is independent of temperature if $\mathbf{m}$ is a constant.

If the assumption is no longer made that the dipole moment $\mathbf{m}$ of a single unit is constant, then similar conclusions to those above can be drawn from the temperature-dependence of $B(T)$. A decrease of $B(T)$ as the temperature is lowered can now mean either an increasing tendency towards anti-parallel orientation between the dipoles of neighbouring units (molecules or unit cells) or a decrease of the dipole moment of a single unit, or vice versa if $B(T)$ increases.

In Figs. 38(i) and (ii) a few examples of this analysis are given. We shall now show how in the case of a known structure the general theory can be applied.

Water [K4, O2, K6]

Water no doubt is one of the most important dielectrics, and the correct calculation of its dielectric constant must be considered as a great success of Kirkwood's [K4] formula 8.5, 8.14 which applies to dipolar liquids in general. A water molecule can be considered as a negative O^{--} ion with the two positive H^+ ions attached in such a way that the lines connecting the latter with the centre of the oxygen ion form an angle of 105°, as can be concluded from the infra-red absorption bands of H_2O vapour. The dipole moment of the H_2O molecule is a vector directed along the line dividing the H—O—H bond angle into two equal angles of $52\frac{1}{2}°$. In the liquid state the position of an individual molecule is strongly correlated with those of its neighbours, and X-ray

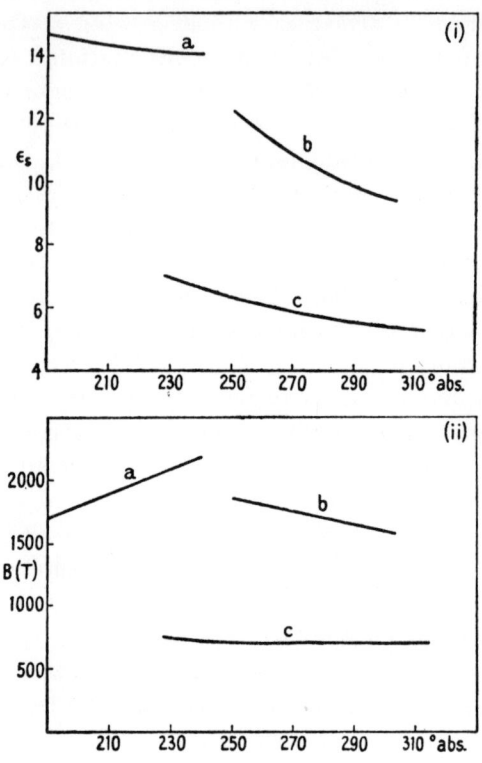

FIG. 38. Examples of $\epsilon_s - T$ analysis. (ii) Temperature-dependence of the function $B(T)$ obtained from $\epsilon_s(T)$ (cf. i) with the help of equation 17.5. (a) Dichloro-propane, $(CH_3)_2CCl_2$, solid, using $\epsilon_s(0) = 2·23$. (b) Tertiary-butyl-chloride, $(CH_3)_3CCl$, liquid, using $\epsilon_s(0) = 2·45$. (c) Penta methyl-chlor-benzene, $(CH_3)_5Cl$, solid, using $\epsilon_s(0) = 2·8$. Measurements: (a), (b) Turgewich and Smyth [T2]; (c) White, Biggs, and Morgan [W3].

With decreasing temperature, (a) indicates increasing tendency to anti-parallel—(b) to parallel—orientation of dipoles. (c) points to random orientation, and to validity of Onsager's formula.

investigations show that the average number of nearest neighbours is close to four. Also, X-ray investigations of liquids show that in general the nearest neighbours of a molecule are arranged in a fairly ordered way corresponding to a certain type of crystal structure. In contrast to solids, however, this order gradually disappears for more distant molecules. In water, according to the model of Bernal and Fowler [B2], the four nearest neighbours of a given molecule form a nearly regular tetrahedron with that molecule at its centre. The bond between neighbouring molecules, called the hydrogen bond by chemists, is supposed to be directed along the O—H bond of one molecule towards the oxygen ion of the other. The arrangement of a molecule and its four neighbours may be represented schematically as in the diagram, (Fig. 39) in which the hydrogen bonds are indicated by dotted lines.

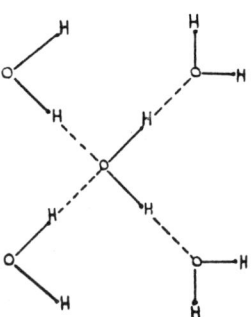

Fig. 39. A H_2O molecule with its four neighbours, schematically. Full lines represent ordinary chemical bond, dotted lines indicate hydrogen bond.

To simplify the following considerations we shall assume the bond angle to be $2\cos^{-1} 1/\sqrt{3} \simeq 109°$ instead of 105°. This should make little difference to the final result. It enables us to consider the tetrahedron formed by the four neighbours of a given molecule as regular, with that molecule at the centre. A simple way to picture the arrangement of molecules and their dipoles is to start with a cube with four of its vertices forming a regular tetrahedron, and the centre occupied by H_2O molecules (Fig. 40). The four bonds emanating from the centre namely two —O—H···O⟨ bonds and two ⟩O···H—O— bonds, follow the lines joining the centre with the vertices. A dipole can have any one of the six directions perpendicular to the faces of the cube. Once the direction of the central dipole is fixed, however, each of the neighbouring dipoles can have one of only three possible directions: for the two molecules bound by —O—H···O⟨ bond the dipole directions are away from our cube, and for the other two towards it, as can be seen from the figure. The other three directions are excluded because they would require —O—H···H—O— bonds. It will now be further assumed that correlation with more distant neighbours need not be considered. Then the three possible dipole directions of the neighbours will be equally probable. Hence the average value $\overline{\cos\gamma}$ of $\cos\gamma$, where γ is the angle between neighbouring dipoles, is

$$\overline{\cos\gamma} = \tfrac{1}{3}, \qquad (17.7)$$

because in two positions the dipoles are perpendicular to each

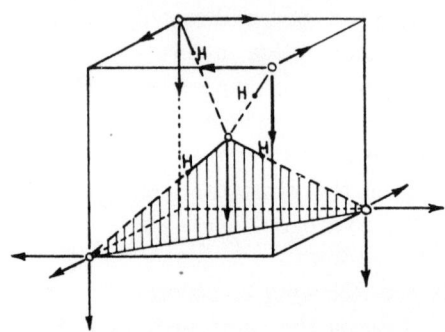

Fig. 40. An H_2O molecule with its four neighbours; dotted lines represent hydrogen bond. The arrows at the corners show the possible dipole directions of the neighbours for a fixed dipole direction of the central molecule.

other ($\cos\gamma = 0$) and in one they are parallel ($\cos\gamma = 1$). The same result can be obtained if the O—H bond outside the centre is permitted to have any direction which does not destroy the O—H$\cdots$O bond; in other words, if free rotation around the O—H$\cdots$O bond is assumed (cf. Fig. 41). For this leads to an average value $\bar{\mu} = \mu_i/\sqrt{3}$ in the direction of the O$\cdots$H—O bond, if μ_i is the dipole moment of a molecule, and hence to a value $\bar{\mu}/\sqrt{3} = \mu_i/3$ in the direction of the fixed dipole.

Upon inserting the value of $\cos\gamma$ from 17.7 into 8.14, using $z = 4$ for the number of neighbours, we find since

$$1 + z\,\overline{\cos\gamma} = 1 + 4/3 = 7/3,$$

$$\epsilon_s - n^2 = \frac{3\epsilon_s}{2\epsilon_s + n^2}\left(\frac{n^2+2}{3}\right)^2 \frac{4\pi}{3}\frac{\mu_v^2 N_0}{kT}\frac{7}{3}, \qquad (17.8)$$

or since the refractive index $n = 1\cdot 33$, i.e. $n^2 \ll \epsilon_s$,

$$3\epsilon_s/(2\epsilon_s + n^2) \simeq 3/2,$$

$$\epsilon_s \simeq \frac{14\pi}{3}\left(\frac{n^2+2}{3}\right)^2 \frac{\mu_v^2 N_0}{kT}. \qquad (17.9)$$

Now substituting the following numerical values: Avogadro's number $= 6 \times 10^{23}$, density $= 1$, molecular weight $= 18$, $k = 1 \cdot 4 \times 10^{-16}$, $\mu_v = 1 \cdot 9 \times 10^{-18}$, $n^2 = 1 \cdot 77$, we find, using 15.2,

$$\epsilon_s \simeq \frac{19000}{T} \quad \text{i.e.} \quad \epsilon_s \simeq 63 \quad \text{for} \quad T = 300° \text{ abs.},$$
(17.10)

which compares favourably with the experimental value of 78.

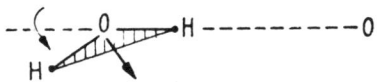

FIG. 41. Direction of the dipole relative to an O—H...O bond.

At 80° C., i.e. at about 353° abs., the experimental value of ϵ_s is 60 and the theoretical one is 53. This shows that the experimentally determined temperature-dependence is also in fair agreement with the $1/T$ law given by 17.10.

In judging the approximations made above it must be remembered that the rigid structure assumed for a molecule and its neighbours does give only an approximate picture of the correlation between neighbours. X-ray investigations [M4] show that the average number of neighbours is slightly larger than four; this may be the reason for the larger experimental value of ϵ_s. Also, with increasing temperature the restriction placed on the relative orientations of neighbouring dipoles by the hydrogen bond may occasionally be overcome. This would decrease the dielectric constant and account for a stronger temperature-dependence than that given by equation 17.10.

It thus seems that on the whole the Kirkwood formula accounts in a satisfactory way for the static dielectric constant of water.

No satisfactory theoretical treatment of the frequency-dependence of the dielectric properties of water is available at present. This would require a generalization of the Kirkwood formula applicable to time-dependent fields. The experimental results by Collie, Hasted, and Ritson [C3] and by Saxton and Lane [S2] point to very simple properties. It seems that the

Debye formulae 10.15–10.17 are satisfied, and moreover that the relaxation time τ is connected with the viscosity η by Debye's relation 11.30. This means that τ and η/T have the same temperature-dependence; hence τ must be a linear function of η/T. Figure 42 shows that this holds for water as well as for heavy water.

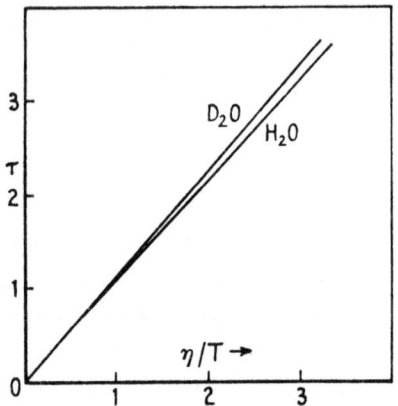

FIG. 42. Dependence of relaxation time τ on η/T (η = viscosity) for water and heavy water, according to Collie, Hasted, and Ritson [C3].

Ketones [F8]

As an example of long-chain substances we shall consider solid ketones, and we shall begin with a discussion of their structure in the lowest energy states which is realized at sufficiently low temperatures. A ketone molecule is obtained from a paraffin molecule by replacing one (or several) CH_2 group(s) by one (or several) CO group(s). In a solid a ketone molecule forms a plane zigzag, as shown in Fig. 28, § 16. The dipole lies in the chain plane and is directed perpendicular to the chain axis. In the simplest case the crystal structure is similar to that of paraffins which has been investigated by Muller [M6]. The molecular chains are arranged in layers whose thickness is approximately equal to the chain length. Within such a layer the molecules form rectangular cells with side length

$$a \simeq 5 \cdot 10^{-8} \text{ cm.}, \quad b \simeq 7 \cdot 5 \times 10^{-8} \text{ cm.},$$

and c slightly larger than the chain length. The dipoles of such a layer are all in one plane, the dipolar plane, and they are arranged as shown in Fig. 43. These dipolar planes are thus polarized in the b-direction, and it is of great importance in studying the behaviour of the whole crystal to know the relative directions of the polarization of successive dipolar planes. First then consider the position of successive layers in paraffins. According to Muller, distinction must be made between two cases, depending upon whether the number of C-atoms in a chain is even or odd. Fig. 44 shows the positions of two successive

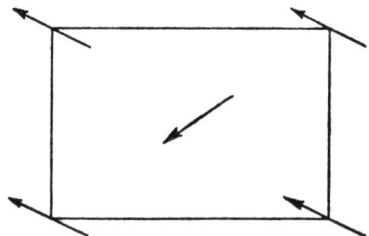

Fig. 43. Dipole directions in the ground state of a dipolar plane.

layers for the two cases. Now let us attach a dipole to each molecule at distances c_1 and c_2 from the two ends so that c_1+c_2 is equal to the chain length. For both even and odd chains there now exist two possibilities. Except for the gap between the chains the distance between successive dipolar planes is (a) c_1+c_2, or (b) it is alternately $2c_1$ and $2c_2$. Figure 45 shows that for odd chains both cases lead to opposite directions of the polarization of successive dipolar planes. For even chains, however, this is only so in case (b), whereas in case (a) successive layers have the same direction of polarization. This latter case will thus lead to a strong polarization of the crystal with all its dipoles nearly parallel, while in the other cases the polarization of successive layers cancel.

For paraffins both cases, (a) and (b), are identical. Their energies, therefore, differ only by the contributions of the dipoles.

The interaction between dipoles can be considered as composed of the interaction between the dipoles of each single layer and of the interaction between dipolar layers. Since the distance between neighbouring dipolar layers is large compared with the distance between neighbouring dipoles within a layer, it is possible to approximate a dipolar plane by a continuously polarized plane. The interaction of a given dipolar plane with all the others is

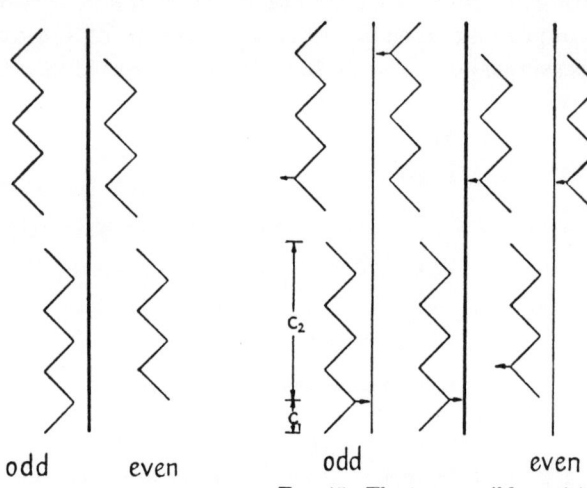

FIG. 44. Position of paraffin chains in successive layers (schematically).

FIG. 45. The two possible positions of dipoles in the ground state of odd and even ketones (schematically).

then equal to its interaction with the surface charge which the other planes produce at the surface of the specimen. For a sufficiently large specimen this interaction energy should be nearly equal to the self-energy of a continuously polarized specimen of the same size and with the same total moment. According to the appendix (A 2.iii) this energy is positive, but its value depends on the shape. It thus follows that the energy of the spontaneously polarized structure is higher than that of the other structures, so that we cannot expect the spontaneous formation of permanently polarized crystals, corresponding to ferro-magnetic substances in the magnetic case. On the other hand, if such a structure could be obtained in an even ketone, it would probably persist for a very long time, because to reach an unpolarized state the chains of some dipolar planes would be required to reverse the direction of their chain axis; this should hardly be feasible in the solid state. It is possible that the required structure can be obtained by solidifying an even ketone in the presence of a strong electric field. The strength of the field required depends on the shape of the specimen; the most favourable case would be that of a needle-shaped specimen with its axis parallel to the field, because such a shape leads to the smallest self-energy.

Now let us consider the dielectric properties of ketones which are not permanently polarized. At low temperature the only effect of the field will be to displace slightly the equilibrium direction of the dipoles, apart from the displacement of electrons. By comparing, at these temperatures, the static dielectric constant of a ketone with that of a paraffin of equal chain length, Muller [M7] found that the contribution of dipole displacement is of the order

$$\Delta \epsilon \simeq 0 \cdot 1. \tag{17.11}$$

This value can be used to estimate the frequency $\omega_0/2\pi$ of the rotational oscillations of the dipoles. Suppose that the field **E** makes an angle θ with the equilibrium direction of a dipole μ. It then exerts a couple $\mu E \sin\theta$ on the dipole. The restoring couple is $I\omega_0^2 \phi$ if I is the moment of inertia and ϕ the average angular displacement of the dipole direction towards the field direction; hence

$$\phi = \frac{\mu E \sin\theta}{I\omega_0^2}. \tag{17.12}$$

For weak fields, $\phi \ll 1$, so that $\mu\phi \sin\theta$ is the projection of the induced dipole in the field direction. Averaging over all direction between μ and **E**, i.e. replacing $\sin^2\theta$ by $\tfrac{2}{3}$, leads to an induced moment per unit volume

$$M_E = \frac{2}{3}\frac{\mu^2 E N_0}{I\omega_0^2} \tag{17.13}$$

if N_0 is the number of dipoles per unit volume. Hence using 1.9 the contribution to the dielectric constant is

$$\Delta\epsilon = \frac{8\pi \mu^2 N_0}{3\ I\omega_0^2}. \tag{17.14}$$

For the ketone used in Muller's experiments $N_0 \simeq 2 \times 10^{21}$ per c.c., $\mu \simeq 2 \cdot 5 \times 10^{-18}$ e.s.u. To obtain I we first assume that the C-atom remains in its initial position and the O-atom can be displaced. Then $I = Ar^2$, where $A = 16 \times 2 \times 10^{-24}$ gramme is the atomic mass of oxygen, and $r \simeq 10^{-8}$ cm. Hence using 17.11 and 17.14, the resonance wave-length λ_0 is of the order

$$\lambda_0 = \frac{2\pi c}{\omega_0} = \frac{2\pi c}{\mu}\left(\frac{3I \Delta\epsilon}{8\pi N_0}\right)^{\frac{1}{2}} \simeq 10^{-2} \text{ cm}. \qquad (17.15)$$

On the other hand, if we were to assume that the chain oscillates rigidly, then I would be larger by a factor of the order of the chain length (~ 20), and hence λ_0 would be larger by about $\sqrt{20}$. Thus we expect a resonance wave-length between 1/10 and 1/100 cm. This may well lead to a measurable absorption in the centimetre region.

It was shown in § 16 (cf. Fig. 30) that for dilute solutions of ketones in a paraffin crystal with longer chain length there are two equilibrium positions for each molecule with opposite dipole direction and about equal energy. In a pure ketone crystal, too, it seems likely that an equilibrium position can be obtained by turning a chain plane by 180° around the chain axis. This new position must, however, have a higher energy than the former one, partly due to the alteration of the interaction between the dipoles and partly due to non-dipolar interaction. The interaction between dipoles is mainly restricted to those of the same dipolar plane. Hence, in view of the strong anisotropy of ketone crystals, calculation of this interaction is reduced to a two-dimensional problem. In this case the interaction with nearest neighbours is a good approximation to the total interaction. Since the non-dipolar interaction is also restricted to nearest neighbours it is not necessary to take into account any long-range interaction if we consider non-polarized states of the crystal. It follows from these considerations that with increasing temperature an increasing number of dipoles will turn in the opposite direction, leading to an order-disorder transition in the way discussed in § 8 and in the first section of the present paragraph. The dielectric constant should, therefore, rise with temperature, as indicated in Fig. 10, until the transition temperature has been reached. Actually Muller's experiments show that the crystal melts before the transition temperature has been reached.† This possibility was envisaged in our previous discussion.

A transition of the type considered here was found, however, by Ubbelohde [$U1$] in a paraffin by measurement of the specific

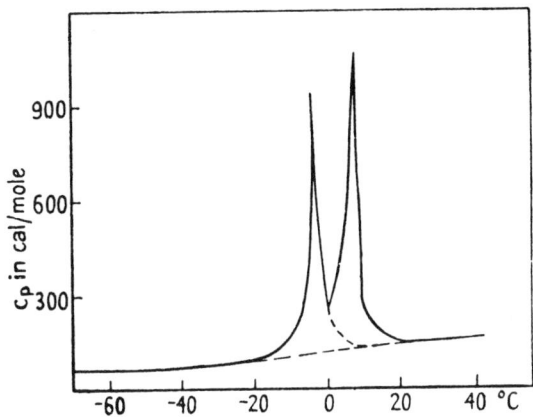

FIG. 46. Temperature-dependence of the specific heat c_p of a paraffin ($m = 15$) according to Ubbelohde [$U1$], showing a transition at about $-3°$ C. and melting. Extrapolations of 'normal' specific heat is indicated.

heat c_p at constant pressure. The turning of molecules into different equilibrium positions makes a contribution Δc_p to c_p which is characteristic of order-disorder transitions. It leads to a sharp increase of c_p near the transition temperature followed by an abrupt fall beyond it. Figure 46 shows that Δc_p can be separated from the normal specific heat with fair accuracy. The transition is connected with a change ΔS in entropy which is given by the well-known thermodynamic relation

$$\Delta S = \int \frac{\Delta c_p}{T} dT \qquad (17.16)$$

integrated over all temperatures. If ΔS is the entropy increase per molecule, then in the disordered state a molecule has $e^{\Delta S/k}$ positions for each one in the ordered state. We should thus expect
$$\Delta S = k \log 2. \qquad (17.17)$$

† This transition has now been found by Dr. Vera Daniel (*Nature*, 1949) in a paraffin-ketone solution. *Added in proof.*

Actually, however, the experimental value of ΔS is $3k \simeq k \log 20$. This means that above the transition temperature a chain must have many more than two positions. It is likely that this is due to twisting of chains [*F6*]. This may be expected in view of the flexibility of chains which has been demonstrated previously in the discussion of the relaxation times of dilute solutions of ketones (§ 16).

Additional evidence of the twisting of chains near the transition temperature has been given by Muller [*M8*]. He measured

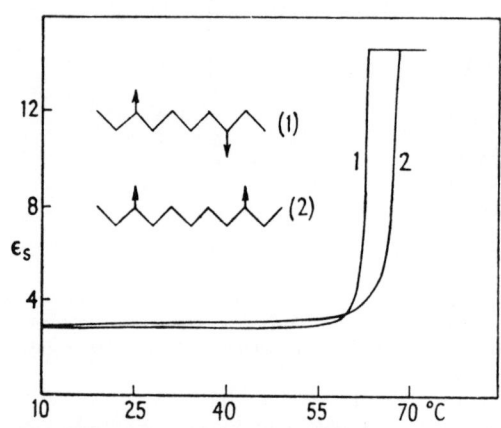

FIG. 47. Temperature-dependence of the dielectric constant of the two diketones $C_{10}H_{18}O_2(1)$ and $C_{11}H_{20}O_2(2)$ according to Muller [*M8*]. The molecular chains in the untwisted state are shown as zigzags, the arrows indicating the directions of the dipolar ketone groups.

the dielectric constant of ketones containing two ketone groups, i.e. two equal dipoles, for two different cases: (i) the number of links between the ketone groups is odd, and (ii) it is even. In case (i) the dipoles of the ketone groups are anti-parallel in the untwisted chain, so that the total dipole moment vanishes whereas in case (ii) they are parallel. Nevertheless, in this case the dielectric constant showed a similar increase below the melting-point to that of case (i) (cf. Fig. 47). This is possible only if chain twisting occurs in this temperature region, for otherwise the ketone in case (i) would behave as a non-polar molecule and have a dielectric constant independent of temperature.

18. Ionic crystals

Ionic crystals are materials in which each lattice point is occupied by an ion. Examples are the alkali halides such as rock-salt, NaCl, which forms a simple cubic lattice consisting of two interlocking face-centred cubic lattices of Na$^+$ and of Cl$^-$ ions. Normally the negative ions are much larger than the positive ones and often the negative charge of the former overlaps into the region of the neighbouring positive ions so that it is not always possible to refer to the charge of an ion as an integral multiple of $\pm e$. In the case of alkali halides the success of Born's [B4] lattice theory suggests, however, that the charges of positive and negative ions are well separated. For in this theory the binding energy and other properties of the crystals are calculated on the assumption that the interaction between ions is composed of (i) the attraction between the ions of charge $\pm e$, (ii) the repulsion between nearest neighbours, and (iii) some corrective terms due to van der Waals forces.

The polarization of ionic crystals is entirely due to elastic displacements, in both the optical and the infra-red regions. Optical polarization, as indicated in § 14, is due to displacement of electrons relative to the nuclei and the corresponding resonance frequencies are in the visible or ultraviolet region. Infra-red polarization is connected with the displacement of nuclei, and usually is accompanied by a displacement of electrons relative to the nuclei. In contrast to its relatively minor part in other substances, infra-red polarization plays an essential role in the case of ionic crystals. We shall, therefore, have to consider it more closely than has been done hitherto. The most general (not necessarily homogeneous) polarization can be considered as a superposition of plane polarization waves whose wave-lengths range from the order of the lattice distance to infinity, the latter corresponding to homogeneous polarization. The frequencies of polarization waves fall into two main regions corresponding to the frequencies of ultra-violet and of infra-red electromagnetic radiation, but the wave-lengths in both regions may range from values of the order of the lattice distance to infinity. Polarization waves whose wave-lengths are smaller than the size of the specimen can be divided into longitudinal and transverse waves, assuming the crystal to be electrically isotropic. For a plane wave moving in the **k** direction with wave-length $2\pi/k$ and frequency $\omega/2\pi$,

$$\mathbf{P} \propto e^{i\mathbf{kr}-i\omega t}, \quad \mathbf{E} \propto e^{i\mathbf{kr}-i\omega t}, \quad \mathbf{D} = \mathbf{E} + 4\pi\mathbf{P} \propto e^{i\mathbf{kr}-i\omega t}, \quad (18.1)$$

where **D** and **E** are the electric displacement and the electric field-strength due to the polarization wave. If a magnetic field is also assumed to be connected with the polarization wave, then with the use of 18.1 the Maxwell equations lead to the well-known formula $k = n\omega/c$ (n = refractive index). Thus if we exclude polarization waves with this wave number, i.e. if we exclude the presence of electromagnetic radiation, then the magnetic field and the magnetic displacement must vanish. Assuming also the absence of free electric charges and of conduction currents, we have $\rho = 0, j = 0, B = 0$, and hence from the Maxwell equations conclude that (cf. A 1.1 and A 1.3) for polarization waves

$$\text{div } \mathbf{D} = 0, \quad \text{curl } \mathbf{E} = 0, \quad (18.2)$$

or using $\mathbf{D} = \mathbf{E} + 4\pi\mathbf{P}$ and 18.1,

$$\text{div } \mathbf{E} = -4\pi \text{ div } \mathbf{P} = -4\pi i \mathbf{k} \mathbf{P}. \quad (18.3)$$

It follows that the three vectors **E**, **P**, and **D** are parallel. Now for longitudinal waves **k** is parallel to **P** or **D**, so that

$$\text{div } \mathbf{D} = i k D = 0$$

requires that $\mathbf{D} = 0$. For transverse waves, however, **k** is perpendicular to **D** or **P**, so that 18.3 yields div $\mathbf{E} = 0$, which together with the second condition 18.2 leads to $\mathbf{E} = 0$. We have thus shown that

$$\mathbf{D} = 0, \text{ i.e. } \mathbf{E} = -4\pi\mathbf{P} \text{ for longitudinal waves}; \quad (18.4)$$

$$\mathbf{E} = 0, \text{ i.e. } \mathbf{D} = 4\pi\mathbf{P} \text{ for transverse waves}. \quad (18.5)$$

Consider now a spherical specimen whose radius is large compared with the lattice distance, but small compared with $c/\nu\epsilon_s^{\frac{1}{2}}$ where ν is the frequency of the applied electric field. Practically we are thus interested in homogeneous—or nearly homogeneous—polarization of the sphere. In this case there is no difference between longitudinal and transverse waves. Moreover, it will be shown presently that the frequencies of these long polarization waves in a sphere are different from those of a specimen which is large compared with the wave-length. For this case of homogeneous polarization, if $\mathbf{P}_o$ and $\mathbf{P}_{ir}$ are, at any instant, the optical and infra-red polarizations respectively (cf. § 14), the total polarization is given by

$$\mathbf{P} = \mathbf{P}_o + \mathbf{P}_{ir}. \tag{18.6}$$

As shown in § 14, the optical polarization P_o is entirely due to the displacement of electrons relative to the atomic nuclei (electronic polarization). The infra-red polarization, however, is due in part to the displacement of whole ions (considered as rigid; atomic polarization) and in part to a displacement of electrons which results directly from this change in position of the ions. This electronic displacement 'induced' by the motion of the ions vanishes only in exceptional cases. There are thus two displacements of the electrons superimposed upon each other. However, since the second one is considered produced by the change in position of the ions it is essentially an infra-red frequency phenomenon and should be lumped with the 'atomic' polarization to form P_{ir}. The two components of the polarization P_o and P_{ir} can then be considered as independent of each other, i.e. they superpose linearly without perturbing each other.† If the total polarization were divided into components in a different manner—for instance 'atomic' and 'electronic' components—these could not be considered independently since the atomic displacement is usually accompanied by a certain amount of electronic displacement.

Assume now that the sphere is brought into a homogeneous field E_0 due to sources outside the sphere. Then in equilibrium, according to A 2.16, the field inside the sphere is

$$\mathbf{E} = \frac{3\mathbf{E}_0}{\epsilon_s + 2}, \quad \text{static case,} \tag{18.7}$$

if E_0 is static. Hence, using 1.9,

$$\mathbf{P} = \frac{3}{4\pi} \frac{\epsilon_s - 1}{\epsilon_s + 2} \mathbf{E}_0, \quad \text{static case.} \tag{18.8}$$

† Vibrations are called normal vibrations if they superpose linearly.

If the external field E_0 is not static, but has a frequency which is large compared with the infra-red resonance frequency but smaller than the optical resonance frequency, then only optical polarization will be excited, and the dielectric constant is $\epsilon = n^2$. Thus corresponding to 18.8 we find

$$\mathbf{P}_o = \frac{3}{4\pi} \frac{n^2-1}{n^2+2} \mathbf{E}_0. \tag{18.9}$$

This equation holds for all frequencies smaller than those within the optical resonance band, including frequencies smaller than the infra-red resonance frequency, although in the latter case the relation $\epsilon = n^2$ is no longer valid.

Now let α, α_o and α_{ir} be the total, the optical, and the infra-red polarizability of the sphere, respectively. Then by definition (§ 14), if V is the volume of the sphere,

$$\mathbf{P} = \frac{\alpha}{V} \mathbf{E}_0, \qquad \mathbf{P}_o = \frac{\alpha_o}{V} \mathbf{E}_0, \qquad \mathbf{P}_{ir} = \frac{\alpha_{ir}}{V} \mathbf{E}_0, \quad \text{static case},$$
$$\tag{18.10}$$

holds for the equilibrium polarizations in a static field. Hence, using 18.9, 18.8, and 18.6, if a_m is the radius of the sphere, i.e. $V = 4\pi a_m^3/3$,

$$\frac{\alpha_{ir}}{a_m^3} = \frac{\epsilon_s-1}{\epsilon_s+2} - \frac{n^2-1}{n^2+2} = \frac{3(\epsilon_s-n^2)}{(\epsilon_s+2)(n^2+2)}. \tag{18.11}$$

We shall now consider vibrations of homogeneous polarization (i.e. polarization waves of infinite wave-length) of a sphere. We shall assume the material to have one infra-red frequency only for this type of polarization and denote by $\omega_s/2\pi$ its value in the absence of an external field. Assuming the vibrations to be harmonic, P_{ir} must thus satisfy the equation $\ddot{P}_{ir} + \omega_s^2 P_{ir} = 0$ in the absence of an external field. Apart from a constant factor, the first term in this equation represents the rate of change of momentum of the ions, and the second term is the restoring force except for its sign. In the presence of an external field E_0, which may depend on time, the right-hand side of the above equation must be equal to the force exerted by this field (apart from the constant factor), i.e. it must be proportional to E_0 where the proportionality factor is independent of the frequency. This factor can, therefore, be obtained by considering the static case where $\ddot{P}_{ir} = 0$ and where the last equation of 18.10 must be fulfilled. Thus

$$\frac{1}{\omega_s^2}\ddot{\mathbf{P}}_{ir}+\mathbf{P}_{ir} = \frac{3}{4\pi}\frac{\alpha_{ir}}{a_m^3}\mathbf{E}_0. \qquad (18.12)$$

We shall now return to the study of polarization waves in a specimen which is large compared with the wave-length. To calculate the infra-red frequency in this case we shall consider, within this specimen, a spherical region whose radius a_m is small compared with the wave-length but large compared with the lattice distance. The polarization within such a sphere is thus nearly homogeneous, so that the equations derived above can be applied. In particular, equation 18.12 will hold, if E_0 now represents the field inside the sphere produced by the polarization of the surroundings. This means that E_0 must be periodic with the same period as the polarization wave, and it may be expected to be proportional to P_{ir} (cf. below, equation 18.15). Now within the sphere such a field E_0 gives rise to an induced optical polarization P_o (proportional to E_0) which can be calculated from 18.9 as well as the infra-red polarization given by 18.12. Thus, assuming that 18.12 gives the same value for P_{ir} as was previously considered to exist in the small isolated sphere with no applied field, there will be superimposed upon this an optical polarization due to the field E_0 produced by the remainder of the specimen when the small sphere is merely a region in that specimen. Hence in a large (compared with the wave-length) specimen an infra-red polarization wave is connected with a polarization $\mathbf{P}_{ir}+\mathbf{P}_o$ in contrast to the case of a small (compared with the wave-length) sphere where a polarization wave is connected with a polarization $\mathbf{P}_{ir}$ only (this was the definition of $\mathbf{P}_{ir}$). This means that the polarization $\mathbf{P}_{ir}+\mathbf{P}_o$ due to an infra-red polarization wave in a large specimen is composed of atomic and electronic polarization in a different way than in a small sphere.† We shall show presently that this composition is

† P_{ir}, which was a normal vibration of the sphere, is no longer a normal vibration of a wave in a large specimen.

different for longitudinal and for transverse waves, and that the frequencies of these two types of waves are different as well.

The external field $\mathbf{E}_0$ may be expected to be proportional to the total polarization $\mathbf{P}$, say

$$\mathbf{E}_0 = q\frac{4\pi}{3}\mathbf{P} = q\frac{4\pi}{3}(\mathbf{P}_o + \mathbf{P}_{ir}), \qquad (18.13)$$

where the constant q will be determined later. Inserting 18.13 into 18.9, and solving for $\mathbf{P}_o$, yields

$$\mathbf{P}_o = \frac{q(n^2-1)}{n^2+2-q(n^2-1)}\mathbf{P}_{ir}. \qquad (18.14)$$

Introducing this again into 18.13 leads to

$$\frac{3}{4\pi}\mathbf{E}_0 = q\mathbf{P}_{ir}\left(1 + \frac{q(n^2-1)}{n^2+2-q(n^2-1)}\right) = q\mathbf{P}_{ir}\frac{n^2+2}{n^2+2-q(n^2-1)}. \qquad (18.15)$$

Hence with the use of 18.11 and 18.15, equation 18.12 becomes

$$\frac{1}{\omega_s^2}\ddot{\mathbf{P}}_{ir} + \mathbf{P}_{ir} = \frac{\epsilon_s - n^2}{\epsilon_s + 2}\frac{3q}{n^2+2-q(n^2-1)}\mathbf{P}_{ir}. \qquad (18.16)$$

This can be written as

$$\frac{1}{\omega^2}\ddot{\mathbf{P}}_{ir} + \mathbf{P}_{ir} = 0, \qquad (18.17)$$

leading to vibrations with a frequency $\omega/2\pi$ given by

$$\frac{\omega^2}{\omega_s^2} = 1 - \frac{\epsilon_s - n^2}{\epsilon_s + 2}\frac{3q}{n^2+2-q(n^2-1)}. \qquad (18.18)$$

To determine q we notice that the macroscopic electric field $\mathbf{E}$ inside the sphere can be considered as composed of the contribution $\mathbf{E}_0$ of the region outside the sphere, and of the self-field $\mathbf{E}_s$ of the sphere. According to A 2.21 the latter is equal to $-4\pi\mathbf{P}/3$, so that

$$\mathbf{E} = \mathbf{E}_0 - \frac{4\pi}{3}\mathbf{P}, \qquad (18.19)$$

or inserting $\mathbf{E}$ from 18.4 and 18.5,

$$\mathbf{E}_0 = \mathbf{E} + \frac{4\pi}{3}\mathbf{P} = \begin{cases} \dfrac{4\pi}{3}\mathbf{P} & \text{for transverse waves} \\ \dfrac{-8\pi}{3}\mathbf{P} & \text{for longitudinal waves.} \end{cases} \qquad (18.20)$$

Comparison with 18.13 shows that

$$q = \begin{array}{l} 1 \text{ for transverse waves} \\ -2 \text{ for longitudinal waves.} \end{array} \quad (18.21)$$

Denoting the angular frequencies of longitudinal and transverse waves by ω_l and ω_t respectively, we find by inserting 18.21 into 18.18,

$$\frac{\omega_t^2}{\omega_s^2} = \frac{n^2+2}{\epsilon_s+2}, \qquad \frac{\omega_l^2}{\omega_s^2} = \frac{\epsilon_s}{n^2}\frac{n^2+2}{\epsilon_s+2}. \quad (18.22)$$

The ratio of longitudinal to transverse frequency is thus given by (cf. references *F11, L4, L5, K2*)

$$\frac{\omega_l}{\omega_t} = \left(\frac{\epsilon_s}{n^2}\right)^{\frac{1}{2}}. \quad (18.23)$$

We thus see that the frequency of longitudinal polarization waves is greater than that of transverse waves with an intermediate value for the frequency of a sphere whose radius is small compared with the wave-length. We also find, inserting 18.21 into 18.14, that the induced optical polarization has opposite sign for the two types of waves. Finally it should be emphasized again that these conclusions hold only for wavelengths which are large compared with the lattice distance.

We shall now proceed to calculate the static dielectric constant in terms of the infra-red frequencies. We shall again consider a sphere homogeneously polarized by a constant external field E_0 leading to a static polarization $\mathbf{P} = \mathbf{P}_o + \mathbf{P}_{ir}$. The change $F - F_0$ in free energy due to the infra-red type of polarization P_{ir} can be considered as composed of a self-energy proportional to P_{ir}^2 due to the elastic displacement, and the interaction energy $-\mathbf{P}_{ir}\mathbf{E}_0 V$. Thus following a procedure similar to that in Appendix A 2.iii we obtain

$$F - F_0 = -\mathbf{P}_{ir}\mathbf{E}_0 V + \tfrac{1}{2}C^2 V P_{ir}^2, \quad (18.24)$$

where the constant γ in that appendix has now been denoted as $\tfrac{1}{2}C^2/V$. In equilibrium F must be a minimum if $\mathbf{P}_{ir}$ is treated as a vector parameter, because $\mathbf{P}_{ir}$ and $\mathbf{P}_o$ are independent of each other in the case of a sphere. Hence, as in the development in the appendix,

$$\mathbf{P}_{ir} = \frac{\mathbf{E}_0}{C^2}. \quad (18.25)$$

Inserting this expression into the last equation of 18.10 and making use of 18.11 we find

$$\frac{\epsilon_s - n^2}{\epsilon_s + 2} = \frac{4\pi}{3} \frac{n^2 + 2}{3} \frac{1}{C^2}. \tag{18.26}$$

Now for elastic displacement a generalized displacement co-ordinate **Q** referring to a single unit cell can always be found so that the self-energy of the sphere is given by

$$\tfrac{1}{2} N_0 V M_{\text{red}} \omega_s^2 Q^2, \tag{18.27}$$

where M_{red} is the reduced mass of the ions and N_0 is the number of unit cells per unit volume. This expression must be equal to $\tfrac{1}{2} C^2 V P_{ir}^2$, so that

$$C^2 = \frac{N_0 M_{\text{red}} \omega_s^2 Q^2}{P_{ir}^2}. \tag{18.28}$$

Furthermore, the polarization $\mathbf{P}_{ir}$ must be proportional to the displacement **Q** and to the number of cells per unit volume. We thus can introduce an effective ionic charge e^* by†

$$\mathbf{P}_{ir} = e^* N_0 \mathbf{Q}. \tag{18.29}$$

Inserting the value of C^2 from 18.28 into 18.26 and making use of 18.29 we thus find‡ (cf. Szigeti, *S13*)

$$\frac{\epsilon_s - n^2}{\epsilon_s + 2} = \frac{4\pi}{3} \frac{n^2 + 2}{3} \frac{e^{*2} N_0}{M_{\text{red}} \omega_s^2}. \tag{18.30}$$

If instead of ω_s we introduce the transverse angular frequency ω_t from 18.22 the above equation becomes

$$\epsilon_s - n^2 = 4\pi \left(\frac{n^2 + 2}{3}\right)^2 \frac{e^{*2} N_0}{M_{\text{red}} \omega_t^2}. \tag{18.31}$$

† It should be remembered here that $\mathbf{P}_{ir}$ was defined as polarization due to the infra-red normal vibrations of a small sphere. The effective charge e^* is therefore a quantity which specifically refers to a sphere.

‡ An alternative way of deriving equation 18.30 would be to make use of the general theory of § 7 according to which (7.21, 7.44)

$$\frac{\epsilon_s - 1}{\epsilon_s + 2} = \frac{4\pi}{3V} \frac{\overline{M_{\text{vac}}^2}}{3kT},$$

where $\overline{M_{\text{vac}}^2}$ is the average square of the spontaneous fluctuation of the moment of a dielectric sphere of volume V in vacuum. $\overline{M_{\text{vac}}^2}$ is in our case of harmonic vibrations the sum of optical and of infra-red terms and is proportional to kT.

In deriving this formula we proceeded on macroscopic lines, except when introducing M_{red} and e^* by means of equations 18.27 and 18.29. Of these two constants the reduced mass can always be obtained from the mass of ions and from a knowledge of the structure with the help of Born's [B4] lattice theory. Calculation of the effective charge e^*, however, would require a detailed knowledge of short-range interactions and of the charge distribution in the lattice. The reason for this different behaviour of mass and charge is due to the fact that the mass is concentrated in the nuclei and contributions by the electrons are negligible, whereas the charge cannot be considered as concentrated at points except for long-distance interaction.

As an example consider a crystal of the NaCl type. Let M^+ and M^- be the masses of the positive and negative ions and let $\mathbf{r}^+$ and $\mathbf{r}^-$ be their displacements. In an oscillation of the sphere corresponding to a polarization wave with infinite wave-length the polarization is homogeneous at any instant. This means that the displacements $\mathbf{r}^+$ and $\mathbf{r}^-$ have the same value in each unit cell, and that $\mathbf{r}^+$ is opposite in direction to $\mathbf{r}^-$. The restoring force must then be proportional to $|\mathbf{r}^+ - \mathbf{r}^-|$ and act in opposite direction on the positive and the negative ions. This suggests the relation

$$\mathbf{Q} = \mathbf{r}^+ - \mathbf{r}^-, \tag{18.32}$$

because then if we put the restoring force as equal to $\pm M_{\text{red}} \omega_s^2 \mathbf{Q}$ we obtain 18.27 as the self-energy of the sphere. This assumption for the restoring force requires

$$M^+ \ddot{\mathbf{r}}^+ + M_{\text{red}} \omega_s^2 \mathbf{Q} = 0, \qquad M^- \ddot{\mathbf{r}}^- - M_{\text{red}} \omega_s^2 \mathbf{Q} = 0. \tag{18.33}$$

Dividing by M^+ and M^- respectively and subtracting the two equations we obtain, using 18.32,

$$\ddot{\mathbf{Q}} + M_{\text{red}} \left(\frac{1}{M^+} + \frac{1}{M^-} \right) \omega_s^2 \mathbf{Q} = 0, \tag{18.34}$$

which leads to oscillations of the required angular frequency ω_s if

$$\frac{1}{M_{\text{red}}} = \frac{1}{M^+} + \frac{1}{M^-}. \tag{18.35}$$

We have thus calculated the reduced mass and now require the effective charge e^*, which has to be obtained by calculating the polarization of the sphere for a given displacement Q according to 18.29. On the assumption that no non-dipolar interaction exists, and that the charges of neighbouring ions do not overlap, it is easy to show that e^* is equal to e, where $\pm e$ is the charge of an ion. The assumption of the absence of non-dipolar forces is entirely fictitious, however. In fact the restoring force is mainly due to short-range repulsion between neighbours. A satisfactory calculation of e^* has not yet been carried out.

We have thus shown that at the present stage of development of the theory all quantities in equation 18.31 except e^* can be obtained from experiment. Use of this equation, therefore, allows a semi-empirical determination of e^*. The following table (Szigeti, *S13*) shows that for alkali halides e^*/e is smaller than unity. This need not be taken as an indication of pronounced overlapping of charges, however. It may also indicate that on setting up a homogeneous polarization by a displacement of the nuclei of ions, an electronic polarization in the opposite direction is induced by the short-range forces.

TABLE

$(\lambda_t = 2\pi c/\omega_t)$

	ϵ_s	n^2	$\lambda_t \times 10^4$ cm.	e^*/e
LiF	9·3	1·92	32·6	0·83
NaF	6·0	1·74	40·6	0·94
NaCl	5·6	2·25	61·1	0·76
NaBr	6·0	2·62	74·7	0·85
NaI	6·6	2·91	85·5	0·71
KCl	4·7	2·13	70·7	0·80
KBr	4·8	2·33	88·3	0·76
KI	4·9	2·69	10·2	0·69
RbCl	5·0	2·19	84·8	0·86
RbBr	5·0	2·33	114	0·88
RbI	5·0	2·63	129·5	0·78
CsCl	7·2	2·60	102	0·88
CsBr	6·5	2·78	134	0·81
TlCl	32	5·10	117	1·11
CuCl	10	3·57	53	1·10
CuBr	8	4·08	57	1·0
MgO	10	2·95	17·3	$2 \times 0·88$
CaO	12	3·28	27·4	$2 \times 0·76$
SrO	13	3·31	47	$2 \times 0·60$

For the oxides a factor 2 is separated from the e^*/e value to indicate that an ideal oxygen ion is doubly charged. We see that $\frac{1}{2}e^*/e$ in this case is somewhat smaller but of the same order as e^*/e for alkali halides. It is also noticeable that the high dielectric constant of TlCl does not lead to an excessive value of e^*/e, but is largely due to the high refractive index. In fact, even in the case of TiO_2, whose dielectric constant is larger than 100, a value $\frac{1}{2}e^*/e \sim 0.7$ is obtained (referring to the oxygen ions), as has been shown by Szigeti [S13]. The high refractive index, together with the high charges of the ions, are mainly responsible for the high dielectric constant.

In the above discussion we have made use of equation 18.31 and not of 18.30 because ω_l and not ω_s can be obtained from experiment. Equation 18.31 is very similar to an equation derived by Born [B4] by an approximate method. This equation does not contain the factor $\left(\dfrac{n^2+2}{3}\right)^2$ and it replaces e^* by the actual ionic charge.

It should be remembered that, in view of the relation 18.22 between ω_l and ω_s, equations 18.31 and 18.30 are equivalent. This raises the question of the possibility of permanent polarization of ionic crystals. For on solving 18.30 for ϵ_s it is found that

$$\epsilon_s \to \infty \quad \text{as} \quad \frac{4\pi}{3}\frac{n^2+2}{3}\frac{e^{*2}N_0}{M_{\text{red}}\omega_s^2} \to 1. \tag{18.36}$$

On the other hand, equation 18.22, for $\epsilon_s \to \infty$, requires $\omega_l \to 0$ (if we exclude $\omega_s \to \infty$), which with 18.31 would also lead to $\epsilon_s \to \infty$. Investigations on these lines should be of importance in view of the properties of crystals like barium titanate. They have not been developed far enough, however, to be included in the present book.

APPENDIX

A 1. ELECTRO-MAGNETIC THEORY

(i) Conservation of energy

Macroscopic electromagnetic theory is based on the Maxwell equations,

$$\text{curl } \mathbf{E} = -\frac{1}{c}\frac{\partial \mathbf{B}}{\partial t} \quad (A\ 1.1), \qquad \text{div } \mathbf{D} = 4\pi\rho \quad (A\ 1.3),$$

$$\text{curl } \mathbf{H} = \frac{1}{c}\frac{\partial \mathbf{D}}{\partial t} + \frac{4\pi}{c}\mathbf{j} \quad (A\ 1.2), \qquad \text{div } \mathbf{B} = 0 \quad (A\ 1.4),$$

where $\mathbf{H}$ and $\mathbf{B}$ are the magnetic field strength and induction respectively, ρ is the density of the true charge, $\mathbf{j}$ the density of the conduction current; for $\mathbf{E}$ and $\mathbf{D}$ cf. § 1. To allow a unique calculation of the field vectors $\mathbf{E}, \mathbf{D}, \mathbf{H}, \mathbf{B}$ from ρ and $\mathbf{j}$, the Maxwell equations must be supplemented by two further relations of the type

$$\mathbf{E} = \mathbf{E}(\mathbf{D}) \quad \text{and} \quad \mathbf{H} = \mathbf{H}(\mathbf{B}). \qquad (A\ 1.5)$$

These relations are not included in the fundamental equations for the electromagnetic field, but are characteristic of the type of material employed. For substances of interest in this book

$$\mathbf{H} = \mathbf{B} \qquad (A\ 1.6)$$

can be assumed to hold to a good approximation.

To introduce the energy law, multiply equation A 1.2 by $\mathbf{E}$, and subtract from it equation A 1.1 multiplied by $\mathbf{H}$. Use

$$\mathbf{H} \text{ curl } \mathbf{E} - \mathbf{E} \text{ curl } \mathbf{H} = \text{div}[\mathbf{E} \times \mathbf{H}].$$

Then with equation A 1.6, after multiplication by $c/4\pi$,

$$\frac{1}{4\pi}\left(\mathbf{E}\frac{\partial \mathbf{D}}{\partial t} + \mathbf{H}\frac{\partial \mathbf{H}}{\partial t}\right) + \frac{c}{4\pi}\text{div}[\mathbf{E} \times \mathbf{H}] + \mathbf{j}\mathbf{E} = 0. \qquad (A\ 1.7)$$

Books on electromagnetic theory discuss in detail the fact that the last term of this equation represents, per unit volume, the rate of conversion of electromagnetic energy into other types of energy (heat, kinetic energy of particles, etc.) in so far as this is connected with conduction currents. $c[\mathbf{E} \times \mathbf{H}]/4\pi$ is the Poynting vector representing the rate of flow of energy; the second term, therefore, gives per unit volume the rate of efflux of electromagnetic energy. Then from conservation of energy it follows that the first term must represent the rate of change of energy content per unit volume, provided that there is no flow of energy of another type (e.g. heat currents).

Now
$$\frac{1}{4\pi}\mathbf{H}\frac{\partial \mathbf{H}}{\partial t} = \frac{\partial}{\partial t}\frac{H^2}{8\pi},$$

so that $H^2/8\pi$ is readily recognized as the density of the magnetic energy.

It is, then, often suggested that

$$\frac{1}{4\pi}\int \mathbf{E}(\mathbf{D})\,d\mathbf{D}$$

should represent the density of the electric energy. The integral is considered to extend from $E = 0$ to the actual value. In fact, if $\mathbf{E} = \mathbf{D}/\epsilon_s$, one finds, for a field-strength E_0,

$$\frac{1}{4\pi}\int_0^{E_0} \mathbf{E}\,d\mathbf{D} = \frac{\epsilon_s}{8\pi}E_0^2 \qquad (A\ 1.8)$$

if ϵ_s can be considered as constant. Such a procedure is, however, possible only if $\mathbf{E}$ is a univalent function of $\mathbf{D}$, for otherwise the integral has no unique meaning. Thus only in this case can the electric energy be defined without further investigation.

In general, the conclusion to be drawn is that

$$\frac{1}{4\pi}\mathbf{E}\,d\mathbf{D} \qquad (A\ 1.9)$$

represents the change of energy density (not necessarily electric energy) connected with a variation of $\mathbf{D}$ by $d\mathbf{D}$, if there is no flow of energy of any other type. It has been shown in § 3 how the total energy can be calculated from this expression.

(ii) Conduction current and energy loss in periodic fields

As in § 2, assume a homogeneous field with $E = E_0 \cos\omega t$, and

$$D = \epsilon_1 E_0 \cos\omega t + \epsilon_2 E_0 \sin\omega t = \epsilon_1 E_0 \cos\omega t - \frac{\epsilon_2}{\omega}\frac{\partial E}{\partial t}.$$

Then using $\partial^2 E/\partial t^2 = -\omega^2 E$, and assuming the conduction current j to vanish, the right-hand side of equation A 1.2 becomes

$$\frac{1}{c}\frac{\partial D}{\partial t} = \frac{\epsilon_1}{c}\frac{\partial E}{\partial t} + \frac{\epsilon_2 \omega}{c}E. \qquad (A\ 1.10)$$

If, on the other hand, $j \neq 0$, and if Ohm's law holds, i.e. if (σ is the conductivity)

$$j = \sigma E,$$

and if furthermore $\epsilon_2 = 0$, then the right-hand side of equation A 1.2 reads

$$\frac{\epsilon_1}{c}\frac{\partial E}{\partial t} + \frac{4\pi}{c}\sigma E. \qquad (A\ 1.11)$$

Comparing equations A 1.10 and A 1.11 it follows that in periodic fields the introduction of two dielectric constants $\epsilon_1(\omega)$ and $\epsilon_2(\omega)$ (cf. § 2, equation 2.8) is equivalent to using a single $\epsilon_1(\omega)$, putting $\epsilon_2 = 0$, but introducing instead a frequency dependent conductivity $\sigma(\omega)$. The two representations are connected by

$$\sigma(\omega) = \frac{\omega \epsilon_2(\omega)}{4\pi}. \qquad (A\ 1.12)$$

Joule's law for the rate L of loss of energy per unit volume then becomes

$$L = \sigma \overline{E^2} = \tfrac{1}{2}\sigma E_0^2, \tag{A 1.13}$$

where the bar has been used to represent the average over one period. Inserting σ from A 1.12, this expression is found to be equivalent to equation 3.15.

(iii) Relation between $\epsilon_1(\omega)$ and $\epsilon_2(\omega)$

To derive equations 2.16 and 2.17 apply the theorems of Fourier transformations to equations 2.14 and 2.15.
Then from 2.14,

$$\alpha(x) = \frac{2}{\pi} \int_0^\infty \{\epsilon_1(\mu) - \epsilon_\infty\} \cos \mu x \, d\mu, \tag{A 1.14}$$

and from 2.15,

$$\alpha(x) = \frac{2}{\pi} \int_0^\infty \epsilon_2(\mu) \sin \mu x \, d\mu. \tag{A 1.15}$$

Introducing A 1.15 into 2.14 yields

$$\begin{aligned}
\epsilon_1(\omega) - \epsilon_\infty &= \frac{2}{\pi} \int_0^\infty dx \left(\cos \omega x \int_0^\infty \epsilon_2(\mu) \sin \mu x \, d\mu \right) \\
&= \frac{2}{\pi} \lim_{R \to \infty} \int_0^\infty d\mu \left(\epsilon_2(\mu) \int_0^R \cos \omega x \sin \mu x \, dx \right) \\
&= \frac{2}{\pi} \lim_{R \to \infty} \int_0^\infty \epsilon_2(\mu) \tfrac{1}{2} \left(\frac{1 - \cos(\mu + \omega)R}{\mu + \omega} + \frac{1 - \cos(\mu - \omega)R}{\mu - \omega} \right) d\mu.
\end{aligned} \tag{A 1.16}$$

Now the integrals containing the cos-terms vanish if $R \to \infty$. The remaining terms lead immediately to equation 2.16.

Finally, introducing A 1.14 into 2.15 gives

$$\epsilon_2(\omega) = \frac{2}{\pi} \int_0^\infty dx \left(\sin \omega x \int_0^\infty \{\epsilon_1(\mu) - \epsilon_\infty\} \cos \mu x \, d\mu \right), \tag{A 1.17}$$

which yields equation 2.17 by a similar method.

(iv) Relations between dielectric constants and optical constants

From the Maxwell equations A 1.1–A 1.4 the wave equation can be derived. For this purpose apply the operator curl to A 1.1 and $1/c \, \partial/\partial t$ to A 1.2. Then, using A 1.6, H can be eliminated. Assuming periodic solutions, we have $\mathbf{D} = \epsilon(\omega)\mathbf{E}$, and hence in the absence of free charges (i.e. $\rho = 0$) and of conduction currents ($j = 0$) we find

$$\nabla^2 \mathbf{E} - \frac{\epsilon}{c^2} \frac{\partial^2 \mathbf{E}}{\partial t^2} = 0 \tag{A 1.18}$$

if use is made of the relation (since from A 1.3 and $\rho = 0$, div $\mathbf{E} = 0$)

$$\text{curl curl } \mathbf{E} = \text{grad div } \mathbf{E} - \nabla^2 \mathbf{E} = -\nabla^2 \mathbf{E}.$$

Now assume a solution representing a wave penetrating the dielectric in the x-direction,

$$\mathbf{E} = \mathbf{A} e^{-(\kappa - in)(\omega/c)x - i\omega t}, \tag{A 1.19}$$

where $\mathbf{A}$ is a constant vector perpendicular to the x-direction. Here by the usual definition of the optical constants n is the refractive index and κ is the absorption coefficient. They can be expressed in terms of the complex dielectric constant ϵ (cf. 2.8), for after inserting A 1.19 into A 1.18 we find

$$(n + i\kappa)^2 = \epsilon = \epsilon_1 + i\epsilon_2.$$

Hence
$$\epsilon_1 = n^2 - \kappa^2 \tag{A 1.20}$$

$$\epsilon_2 = 2n\kappa. \tag{A 1.21}$$

Alternatively, using A 1.12,

$$n\kappa = \frac{2\pi}{\omega} \sigma. \tag{A 1.22}$$

A 2. DIPOLE MOMENTS AND OTHER ELECTROSTATIC PROBLEMS

(i) The basic problem

We consider an infinite homogeneous dielectric with static dielectric† constant ϵ_1, which contains a spherical region of radius a and dielectric constant ϵ_2. We wish to calculate the electric field due to any of the following sources:

(a) Outside sources leading to a constant field $\mathbf{E}_\infty$ (say in the z-direction) at a large distance from the sphere.
(b) A point dipole $\mathbf{\mu}$ (say in the z-direction) at the centre of the sphere.
(c) An extended dipole $\mathbf{M}$ (say in the z-direction) inside the sphere arising from a homogeneously polarized sphere of radius a,

$$\mathbf{M} = \frac{4\pi}{3} a^3 \mathbf{P}_c, \tag{A 2.1}$$

$\mathbf{P}_c$ being the polarization.

Let Φ be the electrostatic potential so that

$$\mathbf{E} = -\text{grad } \Phi \tag{A 2.2}$$

is the field-strength. Φ satisfies the Laplace equation

$$\nabla^2 \Phi = 0 \tag{A 2.3}$$

† In contrast to the denotation in the main sections, ϵ_1 and ϵ_2 now represent static dielectric constants.

subject to the following conditions: if r is the distance from the centre of the sphere and θ the angle between $\mathbf{r}$ and the z-axis, then in

case (a) $\quad\quad \mathbf{E} = \mathbf{E}_\infty$, i.e. $\Phi = -E_\infty r \cos\theta$ if $r \gg a$, $\quad\quad$ (A 2.4)

case (b) $\quad\quad \Phi = \dfrac{\mu}{\epsilon_2} \dfrac{\cos\theta}{r^2}$ if $r \to 0$, $\quad\quad$ (A 2.5)

case (c) $\quad\quad \mathbf{D} = \epsilon_2 \mathbf{E} + 4\pi \mathbf{P}_c$ if $r < a$, $\quad\quad$ (A 2.6)

$\quad\quad\quad\quad \mathbf{D} = \epsilon_1 \mathbf{E}$ if $r > a$, $\quad\quad$ (A 2.7)

where $\mathbf{D}$ is the electric displacement. In all three cases the normal component D_r of $\mathbf{D}$ and the tangential component E_θ of $\mathbf{E}$ must be continuous at $r = a$.

The calculations are simplified by treating all three cases together because they all lead to the same angular dependence of Φ. In general theory Φ is developed into a series of spherical harmonics. Owing to our three conditions, and to the boundary conditions, only terms proportional to $\cos\theta$ appear. For this angular dependence the general solution of A 2.3 is given by

$$\Phi = -\left(\frac{A}{r^2} + Br\right)\cos\theta,$$

containing two arbitrary constants A, B. They will have different values outside (A_1, B_1) and inside (A_2, B_2) the sphere and have to be determined from the boundary conditions. Using A 2.4 it follows that $B_2 = E_\infty$, and hence

$$\Phi = -\left(\frac{A_1}{r^2} + E_\infty r\right)\cos\theta, \quad r > a. \quad\quad \text{(A 2.8)}$$

Also using A 2.5, we have $-A_1 = \mu/\epsilon_2$, i.e.

$$\Phi = \left(\frac{\mu}{\epsilon_2 r^2} - B_2 r\right)\cos\theta, \quad r < a. \quad\quad \text{(A 2.9)}$$

According to the boundary conditions

$$E_\theta = -\frac{1}{r}\frac{\partial \Phi}{\partial \theta}$$

must be continuous at $r = a$, so that

$$\frac{A_1}{a^3} + E_\infty = -\frac{\mu}{\epsilon_2 a^3} + B_2. \quad\quad \text{(A 2.10)}$$

Similarly, using A 2.6 and A 2.7, continuity of $D_r = D\cos\theta$ at $r = a$ yields

$$-2\epsilon_1 \frac{A_1}{a^3} + \epsilon_1 E_\infty = \frac{2\mu}{a^3} + \epsilon_2 B_2 + 4\pi P_c. \quad\quad \text{(A 2.11)}$$

From these two equations one finds

$$\frac{A_1}{a^3} = \frac{\epsilon_1 - \epsilon_2}{2\epsilon_1 + \epsilon_2} E_\infty - \frac{3}{2\epsilon_1 + \epsilon_2}\frac{\mu}{a^3} - \frac{4\pi P_c}{2\epsilon_1 + \epsilon_2}, \quad\quad \text{(A 2.12)}$$

$$B_2 = \frac{3\epsilon_1}{2\epsilon_1 + \epsilon_2} E_\infty + \frac{2}{\epsilon_2}\frac{\epsilon_1 - \epsilon_2}{2\epsilon_1 + \epsilon_2}\frac{\mu}{a^3} - \frac{4\pi P_c}{2\epsilon_1 + \epsilon_2}. \quad\quad \text{(A 2.13)}$$

DIPOLE MOMENTS AND ELECTROSTATIC PROBLEMS 165

Inserting A 2.12 and A 2.13 into A 2.8 and A 2.9 gives the solution of the general problem.

For the discussion we shall now separate the three cases.

Case (a): No internal sources, $\mu = 0$, $P_c = 0$. In this case the field inside the sphere is equal to B_2 and will be denoted as cavity field G'. Thus from A 2.13 with A 2.2

$$\mathbf{G'} = \frac{3\epsilon_1}{2\epsilon_1 + \epsilon_2} \mathbf{E}_\infty \qquad (A\ 2.14)$$

is the field inside the sphere. If in particular $\epsilon_2 = 1$, i.e. for an empty sphere

$$\mathbf{G'}(\epsilon_2 = 1) = \mathbf{G} = \frac{3\epsilon_1}{2\epsilon_1 + 1} \mathbf{E}_\infty. \qquad (A\ 2.15)$$

If, on the other hand, the sphere is in vacuum, then $\epsilon_1 = 1$, and the field inside becomes

$$\frac{3}{\epsilon_2 + 2} \mathbf{E}_\infty. \qquad (A\ 2.16)$$

The field outside the sphere according to A 2.8 is composed of the field $\mathbf{E}_\infty$ at infinity and of a dipolar field with potential (using A 2.12)

$$-\frac{\epsilon_1 - \epsilon_2}{2\epsilon_1 + \epsilon_2} E_\infty a^3 \frac{\cos\theta}{r^2}. \qquad (A\ 2.17)$$

Case (b): Point dipole, no external field, $\mathbf{E}_\infty = 0$, $P_c = 0$. In this case, according to A 2.9 and A 2.2, B_2 represents the deviation of the field inside the sphere from a purely dipolar field, i.e. B_2 is the reaction-field R' acting on μ. Hence with A 2.13

$$\mathbf{R'} = \frac{2}{\epsilon_2} \frac{\epsilon_1 - \epsilon_2}{2\epsilon_1 + \epsilon_2} \frac{\mu}{a^3}; \qquad (A\ 2.18)$$

if in particular $\epsilon_2 = 1$, then the reaction field will be denoted by $\mathbf{R}$, and

$$\mathbf{R} = g\mu, \quad \text{where } g = \frac{\epsilon_1 - 1}{2\epsilon_1 + 1} \frac{2}{a^3}. \qquad (A\ 2.19)$$

Outside the sphere we have a dipolar field with potential

$$\Phi = -\frac{A_1 \cos\theta}{r^2} = \frac{3}{2\epsilon_1 + \epsilon_2} \frac{\mu \cos\theta}{r^2}. \qquad (A\ 2.20)$$

Case (c): Homogeneously polarized sphere, no external field, $\mathbf{E}_\infty = 0$, $\mu = 0$. Consider first the sphere in vacuum, i.e. $\epsilon_1 = \epsilon_2 = 1$. Then B_2 represents the self-field inside the sphere which using A 2.13 and A 2.1 is thus given by

$$\mathbf{E}_s = -\frac{4\pi}{3} \mathbf{P}_c = -\frac{\mathbf{M}}{a^3}. \qquad (A\ 2.21)$$

Now surround this sphere by a medium with dielectric constant ϵ_1. Then the increase of the field inside the sphere, $B_2 - E_s$, is the reaction-field R.

Thus using A 2.13 with $\epsilon_2 = 1$,

$$\mathbf{R} = -\frac{4\pi \mathbf{P}_c}{2\epsilon_1+1} + \frac{4\pi}{3}\mathbf{P}_c = \frac{2(\epsilon_1-1)}{2\epsilon_1+1}\frac{4\pi}{3}\mathbf{P}_c = \frac{2(\epsilon_1-1)}{2\epsilon_1+1}\frac{\mathbf{M}}{a^3} \quad (A\ 2.22)$$

in formal agreement with A 2.19.

Thus, if the sphere contains a point dipole μ surrounded by a homogeneously polarized sphere with moment $\mathbf{M}$, then its reaction field is given by

$$\mathbf{R} = \frac{2}{a^3}\frac{\epsilon_1-1}{2\epsilon_1+1}(\mathbf{M}+\mathbf{\mu}) = g(\mathbf{M}+\mathbf{\mu}), \quad (A\ 2.23)$$

where g is defined in A 2.19 and $\mathbf{M}+\mathbf{\mu}$ is the total dipole moment of the sphere.

For regions outside of the sphere also case (b) turns into case (c) if μ is replaced by $\mathbf{M}$. This holds even if $\epsilon_2 \neq 1$. Thus as far as the field in the region external to the sphere is concerned, cases (b) and (c) lead to the same results.

(ii) Dipole moments

The potential of a rigid (non-polarizable) dipole with moment μ_e in an infinite medium of dielectric constant ϵ_1 is given by

$$\Phi = \frac{\mu_e}{\epsilon_1}\frac{\cos\theta}{r^2}. \quad (A\ 2.24)$$

Now let us consider a model of a molecule consisting of a rigid dipole μ at the centre of a sphere with dielectric constant ϵ_2. (The same model will of course apply to an extended dipole with the same moment corresponding to case (c).) If this molecule is embedded in a medium of dielectric constant ϵ_1, then according to A 2.20 (and to the remarks at the end of the discussion of case (c)) the potential outside of the sphere can be represented by equation A 2.24 if

$$\mu_e = \frac{3\epsilon_1}{2\epsilon_1+\epsilon_2}\mu. \quad (A\ 2.25)$$

Therefore μ_e is defined as the external moment of the molecule in the medium ϵ_1.

It should be realized that the moment of the molecule in vacuum, μ_v, is different from μ. For μ_v is defined as the external moment in a medium for which $\epsilon_1 = 1$. Thus the vacuum moment is given by

$$\mu_v = \frac{3}{\epsilon_2+2}\mu. \quad (A\ 2.26)$$

It follows that the external moment μ_e in a medium ϵ_1 can be expressed in terms of μ_v. Using A 2.25 and A 2.26,

$$\mu_e = \frac{\epsilon_2+2}{3}\frac{3\epsilon_1}{2\epsilon_1+\epsilon_2}\mu_v. \quad (A\ 2.27)$$

In contrast to the external moment μ_e—which is the moment a point dipole must have to produce the same field as the molecule—the internal

DIPOLE MOMENTS AND ELECTROSTATIC PROBLEMS   167

moment μ_i is the actual moment of the molecule if embedded in the medium ϵ_1. In vacuum ($\epsilon_1 = 1$) the external and internal moment are, of course, equal. In another medium, the internal moment differs from the vacuum moment μ_v by the polarization of the molecule through the reaction field. Hence equation 6.18 for μ_i is obtained and this, in turn, with the help of A 2.19, leads to equation 6.20. Equation 6.20 can also be derived by using directly the properties of the external moment. As we have seen above, the field produced outside the sphere does not depend on whether the moment is produced by a point dipole or by uniform polarization of the sphere if this leads to the same moment. Therefore the molecule (as represented by a sphere with dielectric constant ϵ_2 and a dipole μ at its centre) must have the same moment as a sphere with dielectric constant ϵ_1 which is in a medium with the same dielectric constant ϵ_1 and has a moment μ_e at its centre, because by definition of μ_e both produce the same field outside the sphere. Thus

$$\mu_i = \mu_e + \int_{\text{sphere}} \mathbf{P}\, d\tau, \qquad (A\ 2.28)$$

showing that μ_i is composed of the moment of the rigid dipole μ_e and the moment contained in a sphere around it. Now if μ_e points in the z-direction, the integral must be a vector with the same direction for reasons of symmetry. Hence using

$$4\pi P_z = (\epsilon_1 - 1)E_z = -(\epsilon_1 - 1)\frac{\partial \Phi}{\partial z} \qquad (A\ 2.29)$$

and

$$\frac{\cos\theta}{r^2} = \frac{z}{r^3} = -\frac{\partial}{\partial z}\left(\frac{1}{r}\right),$$

we find with the help of A 2.24 that

$$\int P_z\, d\tau = \frac{\epsilon_1 - 1}{4\pi}\frac{\mu_e}{\epsilon_1}\int \frac{\partial^2}{\partial z^2}\left(\frac{1}{r}\right) d\tau = \frac{\epsilon_1 - 1}{4\pi}\frac{\mu_e}{\epsilon_1}\frac{1}{3}\int \nabla^2\left(\frac{1}{r}\right) d\tau = -\frac{\epsilon_1 - 1}{3\epsilon_1}\mu_e, \qquad (A\ 2.30)$$

since

$$\int \frac{\partial^2}{\partial x^2}\left(\frac{1}{r}\right) d\tau = \int \frac{\partial^2}{\partial y^2}\left(\frac{1}{r}\right) d\tau = \int \frac{\partial^2}{\partial z^2}\left(\frac{1}{r}\right) d\tau$$

and

$$\int \nabla^2\left(\frac{1}{r}\right) d\tau = -4\pi.$$

Inserting A 2.30 into A 2.28 yields

$$\mu_i = \mu_e\left(1 - \frac{\epsilon_1 - 1}{3\epsilon_1}\right) = \frac{2\epsilon_1 + 1}{3\epsilon_1}\mu_e, \qquad (A\ 2.31)$$

which is identical with 6.20.

The above calculation also shows that the dipole moment contained in a sphere surrounding a dipole is independent of the size of this sphere. Hence the moment contained between two spherical surfaces inside the dielectric vanishes. This holds even if the two spheres are not concentric.†

Thus, if we consider a large sphere inside an infinite dielectric of dielectric constant ϵ_1, and this sphere contains a molecule with internal moment μ_i, then the moment of the large sphere is also μ_i.

This is, however, no longer the case if the large sphere is not embedded in its own medium because then the field outside the molecule is different from the field of a point dipole. By analogy with A 2.26 one should expect the moment in this case to be

$$\mu_s = \frac{3}{\epsilon_1+2}\mu_e, \qquad (A\ 2.32)$$

as is actually confirmed by calculation if the radius of the sphere is large compared with that of the molecule.

We have thus introduced five different dipole moments, μ, μ_v, μ_i, μ_e, μ_s. To prevent confusion we give a brief summary.

μ has a meaning only in terms of the special model used.

μ_v = moment of the molecule in vacuum.

μ_i = moment of the molecule in a medium ϵ_1
 = moment of a sphere (containing the molecule) within an infinite medium ϵ_1.

μ_e = moment of a rigid point dipole producing in a medium ϵ_1 the same dipolar field as the molecule.

μ_s = moment of a dielectric sphere in vacuum containing the molecule.

To summarize the formulae:

$$\mu_v = \frac{3}{\epsilon_2+2}\frac{2\epsilon_1+\epsilon_2}{2\epsilon_1+1}\mu_i = \frac{2\epsilon_1+\epsilon_2}{\epsilon_1(\epsilon_2+2)}\mu_e$$

$$= \frac{\epsilon_1+2}{\epsilon_2+2}\frac{2\epsilon_1+\epsilon_2}{3\epsilon_1}\mu_s = \frac{3}{\epsilon_2+2}\mu. \qquad (A\ 2.33)$$

† The moment in any direction, say with radius vector **s**, is proportional to the integral $\int \partial\Phi/\partial s \, d\tau$ extended over the space between the two spheres with radii r_e and r_i. This integral can be transformed into the difference of two integrals over the surfaces of the two spheres (which are designated as external surface e and internal surface i):

$$r_e^2 \int_e \Phi_e \cos\psi \, d\Omega - r_i^2 \int_i \Phi_i \cos\psi \, d\Omega.$$

Here $d\Omega$ is the element of the solid angle for each sphere, Φ_e and Φ_i are the values of Φ at the two surfaces. ψ is the angle between **s** and the normal to the surfaces pointing away from the centre. Both Φ_e and Φ_i can be developed into spherical harmonics of which only the first representing a dipolar potential of a dipole at the centre of the sphere concerned gives a contribution to the integral. These terms are proportional to $\cos\psi/r_e^2$ and $\cos\psi/r_i^2$ respectively, which means that the two surface integrations just cancel.

(iii) Self-Energy

The free energy F of a dielectric polarized by a field can be considered as composed of three terms: (1) the free energy F_0 of the field and of the dielectric before the latter is brought into the field; (2) the energy of interaction between the polarized dielectric and the field; and (3) the free energy required to polarize the dielectric. The latter will be called the self-energy F_s. If the dielectric is polarized homogeneously, then it will be assumed that $F_s = \gamma M^2$, where $\mathbf{M}$ is the moment of the dielectric. The energy of interaction is equal to $-(\mathbf{ME_0})$ if $\mathbf{E_0}$ is the field (supposed homogeneous) before the dielectric is inserted. Thus

$$F - F_0 = -(\mathbf{ME_0}) + \gamma M^2. \quad (A\ 2.34)$$

In equilibrium F must be a minimum where $\mathbf{M} = (M_x, M_y, M_z)$ is treated as a parameter. Hence from

$$\frac{\partial F}{\partial M_x} = \frac{\partial F}{\partial M_y} = \frac{\partial F}{\partial M_z} = 0, \quad (A\ 2.35)$$

we find
$$2\gamma \mathbf{M} = \mathbf{E_0} \quad \text{or} \quad \gamma = \frac{E_0}{2M}. \quad (A\ 2.36)$$

Hence our assumption $F_s = \gamma M^2$ implies that $\mathbf{M}$ is parallel to $\mathbf{E_0}$. The value of γ depends on the shape of the dielectric. Thus a sphere of radius a brought into a homogeneous field $\mathbf{E_0}$ has a moment

$$\mathbf{M} = a^3 \mathbf{E_0}(\epsilon_s - 1)/(\epsilon_s + 2).$$

Hence
$$\gamma = \frac{1}{2} \frac{\epsilon_s + 2}{\epsilon_s - 1} \frac{1}{a^3}, \qquad F_s = \frac{1}{2} \frac{\epsilon_s + 2}{\epsilon_s - 1} \frac{M^2}{a^3}. \quad (A.\ 2.37)$$

A slab of volume V with surface perpendicular to $\mathbf{E_0}$, on the other hand, has a moment $\mathbf{M} = V\mathbf{E_0}(\epsilon_s - 1)/4\pi\epsilon_s$. In this case

$$\gamma = \frac{2\pi}{V} \frac{\epsilon_s}{\epsilon_s - 1}, \qquad F_s = \frac{\epsilon_s}{\epsilon_s - 1} \frac{2\pi M^2}{V}. \quad (A\ 2.38)$$

In both cases, of course, F_s is proportional to the volume if the polarization M/V is constant.

A 3. THE CLAUSIUS–MOSSOTTI FORMULA

In § 5 an equation for the static dielectric constant ϵ_s was derived (5.13) which holds exactly for the model used in that section. This is usually called the Clausius–Mossotti formula. Later in § 6 it was found that a similar formula holds approximately for liquids of non-polar molecules, but that the approximation required negligible short-range interaction. In § 8 it was pointed out that by neglecting short-range interaction the general theory of § 7 will lead to either the Clausius–Mossotti or to the Onsager formula, depending on whether one considers non-polar or polar spherical molecules. The proof of this will be given at the end of this section.

Discussions as to whether or not the Clausius–Mossotti formula holds

APPENDIX

exactly have formed part of the literature on dielectrics over a period of many years. Controversial conclusions have been drawn, mainly due to a misunderstanding of the significance of this formula. In fact one should distinguish between a macroscopic and a molecular formula. The same mathematical symbols are normally used in the two cases and, as a result, they are often confused with each other. The macroscopic formula is valid exactly, but the molecular formula holds only subject to the conditions mentioned in § 15.

To derive the macroscopic formula, consider a sphere of a continuous isotropic dielectric to be brought into a constant electric field **f**. According to equation A 2.16, the field inside the sphere is then given by

$$\mathbf{E} = \frac{3}{\epsilon_s + 2}\mathbf{f} \qquad (A\,3.1)$$

if ϵ_s is the dielectric constant. Let $\mathbf{M}_E$ be the dipole moment induced in the sphere. Then the polarizability α_m of the sphere will be defined by (the suffix m indicates 'macroscopic')

$$\mathbf{M}_E = \alpha_m \mathbf{f}. \qquad (A\,3.2)$$

On the other hand, in view of 1.9,

$$M_E = \frac{(\epsilon_s - 1)}{4\pi} V E = \frac{\epsilon_s - 1}{3} a_m^3 \, E \qquad (A\,3.3)$$

if V is the volume and a_m is the radius of the sphere. Equating A 3.3 and A 3.2 and introducing f from A 3.1 leads to the Clausius–Mossotti formula

$$\frac{\epsilon_s - 1}{\epsilon_s + 2} = \frac{\alpha_m}{a_m^3}. \qquad (A\,3.4)$$

An alternative way of deriving this formula is to consider a spherical region within a homogeneous dielectric. If a constant field **E** is produced in the dielectric, the part of the field inside the spherical region which is due to sources outside this region (the inner field) is given by expression 5.9 and is therefore identical with our field **f** (A 3.1). The remaining development is then identical with the one given above.

Equation A 3.4 is always correct if the spherical region is sufficiently large, so that the material contained in it may be considered from a macroscopic point of view. From a molecular point of view equation A 3.4 has, however, no significance. To give it such a significance the polarizability α must be expressed in terms of other quantities whose values will not be found experimentally by measurement of the static dielectric constant but by experiments of a different nature.

Equation A 3.4 finds its main application in the case of dielectrics in which polarization is connected with elastic displacement of charges (case (i) of § 4). The most general displacement of the dielectric material in a sphere may then be developed in terms of its normal vibrations (cf. van Vleck, $V3$), and it is then found that the polarizability α is a constant independent of temperature if the density of the substance is kept constant. This procedure is of importance because it shows in a very general way that the assumption of elastic displacement leads to a temperature-independent dielectric constant.

We shall now discuss the 'molecular' Clausius–Mossotti formula. In this case it is assumed that the macroscopic polarizability α_m can be expressed in terms of a molecular polarizability α—which is a constant independent of temperature or density—by

$$\alpha_m = N\alpha \qquad (A\ 3.5)$$

if N is the number of molecules in the macroscopic sphere of radius a_m. Furthermore, if

$$\frac{4\pi}{3} a^3 = \frac{V}{N} = \frac{4\pi}{3} \frac{a_m^3}{N} \qquad (A\ 3.6)$$

is the volume occupied per molecule, insertion of equation A 3.5 and A 3.6 into A 3.4 yields

$$\frac{\epsilon_s - 1}{\epsilon_s + 2} = \frac{\alpha}{a^3}. \qquad (A\ 3.7)$$

This molecular formula has the same mathematical structure as formula A 3.4, but its significance is different. In equation A 3.7 the polarizability α is a property of a single molecule independent of macroscopic parameters. The quantity $3/4\pi a^3$ is equal to the number N_0 of molecules per unit volume which is determined by the molecular weight W and the density d,

$$\frac{3}{4\pi a^3} = N_0 = \frac{d}{W} A, \qquad (A\ 3.8)$$

where A is Avogadro's number. Hence α/a^3 is proportional to the density of the dielectric, which may be varied by altering the external pressure. No such conclusion can be drawn from equation A 3.4, where the macroscopic polarizability α_m may depend on the density in an unspecified way.

The crucial step leading from equation A 3.4 to A 3.7 is contained in the hypothesis A 3.5. One should expect A 3.5 to be correct only in the absence of short-range interaction between molecules, because such an interaction would influence the reaction of a given molecule to the field. Actually in the two cases in which we have derived the molecular Clausius–Mossotti formula (5.13 and 6.34) short-range interaction was absent.

Finally let us derive A 3.7 and the Onsager formula by applying the general theory of § 7 to a model in which short-range forces are entirely absent. Also it will be assumed that the unit cell is spherical and that the average polarization outside the unit cell is equivalent to that of an isotropic continuum polarized by the moment **m** of the unit cell. In this case the average moment **m*** of a sphere (embedded in a larger specimen) containing the unit cell is equal to the moment **m** of the unit cell, as was shown in § 7. Hence the spherical region introduced in § 7 may be taken as equivalent to the unit cell, and equations 7.11 and 7.33 become identical if 7.34 with $N = 1$ is taken into account. Thus, using A 3.8,

$$\epsilon_s - 1 = \frac{3\epsilon_s}{2\epsilon_s + 1} \frac{\overline{m^2}}{kTa^3}, \qquad (A\ 3.9)$$

APPENDIX

where $\overline{m^2}$ is the average square of the spontaneous dipole moment of the spherical cell embedded in its own medium. Assuming first that the unit cell contains a rigid dipole with moment μ, $\mathbf{m} = \boldsymbol{\mu}$, and $\overline{m^2} = \mu^2$, the Onsager formula follows from A 3.9. Consider, on the other hand, a spherical polarizable molecule without a permanent moment. Such a molecule can be represented by an elastically bound charge e. Then if $\mathbf{r}$ is the displacement, the internal energy U_i must be quadratic in $\mathbf{r}$, i.e.

$$U_i = \frac{c}{2} r^2, \tag{A 3.10}$$

where c is a molecular constant. To express this constant in terms of the polarizability, assume the molecule to be brought into an external field $\mathbf{f}$. Then its potential energy is given by

$$\frac{c}{2} r^2 - e\mathbf{f}\mathbf{r}. \tag{A 3.11}$$

The equilibrium value $\mathbf{\bar{r}}$ of $\mathbf{r}$ is obtained by making this expression a minimum. Hence

$$e\mathbf{\bar{r}} = \frac{e^2}{c} \mathbf{f} \tag{A 3.12}$$

is the average moment of the molecule. Since this must be equal to $\alpha\mathbf{f}$, we obtain

$$\alpha = \frac{e^2}{c}, \tag{A 3.13}$$

and hence with A 3.10

$$U_i = \frac{e^2}{2\alpha} r^2. \tag{A 3.14}$$

To obtain $\overline{m^2}$ we require the energy $U = U_i + U_e$ because, as in 7.12, using $\mathbf{m} = e\mathbf{r}$,

$$\overline{m^2} = e^2 \int_0^\infty r^2 e^{-U/kT} r^2 \, dr \bigg/ \int_0^\infty e^{-U/kT} r^2 \, dr. \tag{A 3.15}$$

Now the external energy U_e is given by $-\tfrac{1}{2}\mathbf{m}\mathbf{R}$ where $\mathbf{R}$ is the reaction field, as can be shown by the argument which led to 7.18. Therefore making use of 5.10 and the fact that $4\pi a^3/3 = V$,

$$U_e = -\frac{m^2}{a^3} \frac{\epsilon_s - 1}{2\epsilon_s + 1} = -\frac{e^2}{a^3} \frac{\epsilon_s - 1}{2\epsilon_s + 1} r^2. \tag{A 3.16}$$

Introducing $U = U_i + U_e$ into A 3.15 and integrating yields

$$\overline{m^2} = \frac{3}{2} \frac{kT \, 2a^3(2\epsilon_s + 1)\alpha}{a^3(2\epsilon_s + 1) - 2(\epsilon_s - 1)\alpha}. \tag{A 3.17}$$

Inserting this expression into A 3.9 leads at once to equation A 3.7.

In the above derivation the importance of the external energy U_e including the interaction with the reaction field should be stressed

It is interesting to compare this with the case of a non-polarizable molecule with dipole moment μ (Onsager case), where the interaction with the reaction field is irrelevant. In this latter case the only variable is the direction of **μ**, but its amount, μ, is a constant. Thus since **m** = **μ** in this case, $m^2 = \mu^2 = $ constant, so that U_e according to A 3.16 is a constant as well. Hence in A 3.15 the terms containing U_e can be taken outside the integrals and, therefore, cancel.

A 4. SHAPE OF ABSORPTION LINES (*F9*)

We shall consider a set of linear harmonic oscillators of mass M, charge e, proper frequency $\omega_0/2\pi$. Their number N per unit volume is assumed to be so small that interaction between them can be neglected. Then in an external field E, which may depend on time,

$$\ddot{x} = -\omega_0^2 x + \frac{e}{M} E \cos\theta \qquad (A\ 4.1)$$

if x is the displacement of the charge e, θ the angle of **E** with the x-direction, and a dot represents differentiation with respect to time t. The oscillators are supposed to make frequent collisions with a medium which is in thermal equilibrium so that they also tend to reach equilibrium. To describe their behaviour we introduce a distribution function $f(x,\dot{x})$ so that $f(x,\dot{x})\,dxd\dot{x}$ represents per unit volume the number of oscillators with a displacement between x and $x+dx$, and a velocity between $\dot{x}$ and $\dot{x}+d\dot{x}$. Hence

$$\int\!\!\int_{-\infty}^{\infty} f(x,\dot{x})\,dxd\dot{x} = N. \qquad (A\ 4.2)$$

With the help of the distribution function the polarization P in the field direction is obtained as

$$P = e\cos\theta \int\!\!\int_{-\infty}^{\infty} xf(x,\dot{x})\,dxd\dot{x}. \qquad (A\ 4.3)$$

Let us first consider the static case for which the field is independent of time, $E = E_0$. If U is the energy of an oscillator, i.e.

$$U = U_0 - eEx\cos\theta, \qquad (A\ 4.4)$$
$$U_0 = \tfrac{1}{2}M\omega_0^2 x^2 + \tfrac{1}{2}M\dot{x}^2, \qquad (A\ 4.5)$$

then, in equilibrium, according to the Boltzmann theorem,

$$f(x,\dot{x}) = Ce^{-U(x,\dot{x})/kT}, \qquad (A\ 4.6)$$

where C is independent of x and $\dot{x}$ and can be determined with the use of equation A 4.2. Now using A 4.4 and $E = E_0$,

$$e^{-U(x,\dot{x})/kT} = e^{-U_0/kT}\left\{1 - \frac{eEx\cos\theta}{kT} + \frac{1}{2}\left(\frac{eEx\cos\theta}{kT}\right)^2 + \ldots\right\}, \qquad (A\ 4.7)$$

and hence we can write, considering A 4.5,

$$f = f_0 - \frac{eE\cos\theta}{M\omega_0^2}\frac{\partial f_0}{\partial x} + \ldots, \qquad (A\ 4.8)$$

where the dots represent terms proportional to higher powers in E_0. In the calculation of the polarization P we shall be interested in linear

174 APPENDIX

terms in E only. Therefore we need not consider these higher terms; to be exact we should also prove that the series for P, in terms of E_0 converges, but we shall omit this. In A 4.8

$$f_0 = Ce^{-U_0/kT} \qquad (A\ 4.9)$$

represents the equilibrium distribution in the absence of a field. The factor C is the same as in A 4.6 if in calculating the integral A 4.2 for the latter case only terms which are linear in, or independent of, E_0 are considered.

Now inserting A 4.8 into A 4.3 we find for the polarization P_0 in a constant field

$$P_0 = e\cos\theta\left(\iint_{-\infty}^{\infty} xf_0\,dxd\dot{x} - \frac{eE\cos\theta}{M\omega_0^2}\iint_{-\infty}^{\infty} x\frac{\partial f_0}{\partial x}\,dxd\dot{x}\right). \qquad (A\ 4.10)$$

Here the first term vanishes because f_0 is an even function in x. Integrating the second integral by parts we find, using A 4.2,

$$P_0 = \frac{e^2 E \cos^2\theta\, N}{M\omega_0^2}. \qquad (A\ 4.11)$$

Averaging over all directions of the field leads, of course, to a replacement of $\cos^2\theta$ by $1/3$. Hence the contribution $\Delta\epsilon$ to the static dielectric constant is given by

$$\Delta\epsilon = \frac{P_0}{E} = \frac{e^2 N}{3M\omega_0^2}. \qquad (A\ 4.12)$$

Consider now the case of a periodic field E described by

$$E = E_0 e^{-i\omega t}. \qquad (A\ 4.13)$$

Again we shall introduce the distribution function f satisfying A 4.2, but now depending on time t. We shall calculate f by considering its rate of change $\partial f/\partial t$ for a given value of x and $\dot{x}$. This is composed of two terms, one due to collisions of the oscillators with the surrounding medium and the other due to their motion. Denoting by $(\partial f/\partial t)_{\text{coll}}$ the rate of change of f due to collisions the simplest assumption we can make is to put

$$\left(\frac{\partial f}{\partial t}\right)_{\text{coll}} = -\frac{1}{\tau}(f - f_{\text{equ}}), \qquad (A\ 4.14)$$

in which the relaxation time τ is assumed to be independent of x and $\dot{x}$. This implies that equilibrium (f_{equ}) is approached exponentially.

For time-dependent fields we shall assume that A 4.14 still holds if f_{equ} is the equilibrium distribution corresponding to the field at the time in question. Thus making use of A 4.8 and A 4.13,

$$f_{\text{equ}} = f_0 - \frac{eE_0 e^{-i\omega t}\cos\theta}{M\omega_0^2}\frac{\partial f_0}{\partial x}. \qquad (A\ 4.15)$$

For the distribution function f in this case we shall assume

$$f(x,\dot{x},t) = f_0(x,\dot{x}) + g(x,\dot{x})E_0\cos\theta\, e^{-i\omega t}, \qquad (A\ 4.16)$$

where $g(x, \dot{x})$ is a function of x and $\dot{x}$, independent of t. To satisfy A 4.2 this requires

$$\int\int_{-\infty}^{\infty} g(x,\dot{x})dxd\dot{x} = 0. \qquad (A\ 4.17)$$

Inserting A 4.16 and A 4.15 into A 4.14 now leads to

$$\left(\frac{\partial f}{\partial t}\right)_{\text{coll}} = -\frac{1}{\tau}\left(g + \frac{e}{M\omega_0^2}\frac{\partial f_0}{\partial x}\right)E_0 e^{-i\omega t}\cos\theta. \qquad (A\ 4.18)$$

There is a second contribution to $\partial f/\partial t$ due to the motion of the oscillators. This contribution will be denoted by $(\partial f/\partial t)_m$, so that

$$\frac{\partial f}{\partial t} = \left(\frac{\partial f}{\partial t}\right)_{\text{coll}} + \left(\frac{\partial f}{\partial t}\right)_m. \qquad (A\ 4.19)$$

To find an expression for $(\partial f/\partial t)_m$ we notice that all the oscillators contained, at a time t, in an interval $\Delta x \Delta \dot{x}$, the number of such oscillators being $f(x, \dot{x}, t) \Delta x \Delta \dot{x}$, were at the time $t - \delta t$ in an interval of the same size† $\Delta x \Delta \dot{x}$ with coordinates $x - \dot{x}\,\delta t$, $\dot{x} - \ddot{x}\,\delta t$ provided the number has not been changed by collisions; all $f(x, \dot{x}, t - \delta t) \Delta x \Delta \dot{x}$ oscillators contained in this interval at the time $t - \delta t$ have meanwhile passed into another interval. Thus

$$\frac{1}{\delta t}\{f(x - \dot{x}\,\delta t, \dot{x} - \ddot{x}\,\delta t, t - \delta t) - f(x, \dot{x}, t - \delta t)\}$$

is the rate of change of f due to motion. Hence with $\delta t \to 0$,

$$\left(\frac{\partial f}{\partial t}\right)_m = -\frac{\partial f}{\partial x}\dot{x} - \frac{\partial f}{\partial \dot{x}}\ddot{x}, \qquad (A\ 4.20)$$

or making use of A 4.1,

$$\left(\frac{\partial f}{\partial t}\right)_m = -\frac{\partial f}{\partial x}\dot{x} + \omega_0^2\frac{\partial f}{\partial \dot{x}}x - \frac{eE_0\cos\theta\, e^{-i\omega t}}{M}\frac{\partial f}{\partial \dot{x}}. \qquad (A\ 4.21)$$

Now from A 4.9 and A 4.5 it follows that

$$-\frac{\partial f_0}{\partial x}\dot{x} + \omega_0^2\frac{\partial f_0}{\partial \dot{x}}x = 0. \qquad (A\ 4.22)$$

Thus inserting from A 4.16 into A 4.21 and neglecting terms proportional to E_0, we obtain

$$\left(\frac{\partial f}{\partial t}\right)_m = \left(-\frac{\partial g}{\partial x}\dot{x} + \omega_0^2\frac{\partial g}{\partial \dot{x}}x - \frac{e}{M}\frac{\partial f_0}{\partial \dot{x}}\right)E_0 e^{-i\omega t}\cos\theta. \qquad (A\ 4.23)$$

Also from A 4.16, $\qquad \dfrac{\partial f}{\partial t} = -i\omega g E_0 e^{-i\omega t}\cos\theta. \qquad (A\ 4.24)$

Inserting now A 4.24, A 4.23, and A 4.18 into A 4.19 we find that g satisfies the differential equation

$$\left(-i\omega + \frac{1}{\tau}\right)g = -\frac{\partial g}{\partial x}\dot{x} + \omega_0^2\frac{\partial g}{\partial \dot{x}}x - \frac{e}{M}\frac{\partial f_0}{\partial \dot{x}} - \frac{e}{M\omega_0^2\tau}\frac{\partial f_0}{\partial x}. \qquad (A\ 4.25)$$

† This can be proved with the help of A 4.1.

To solve it we put

$$g = a\frac{\partial f_0}{\partial x} + b\frac{\partial f_0}{\partial \dot{x}},\qquad (A\ 4.26)$$

where a and b are constant.

This does not represent the most general solution of A 4.25, but it has been shown by Huby [H3] that it is the only solution satisfying all requirements.

The polarization P according to A 4.3, A 4.16, and A 4.26 is thus given by

$$P = e\cos^2\theta\, E_0\, e^{-i\omega t}\int\!\!\!\int_{-\infty}^{\infty} x\left(a\frac{\partial f_0}{\partial x} + b\frac{\partial f_0}{\partial \dot{x}}\right) dx\,d\dot{x} = -e\cos^2\theta\, E_0\, e^{-i\omega t}\, a. \qquad (A\ 4.27)$$

To find the constant a insert A 4.26 into A 4.25. Making use of

$$\left(-\dot{x}\frac{\partial}{\partial x}+\omega_0^2 x\frac{\partial}{\partial \dot{x}}\right)\frac{\partial f_0}{\partial x} = -\omega_0^2 \frac{\partial f_0}{\partial \dot{x}} \qquad (A\ 4.28)$$

and

$$\left(-\dot{x}\frac{\partial}{\partial x}+\omega_0^2 x\frac{\partial}{\partial \dot{x}}\right)\frac{\partial f_0}{\partial \dot{x}} = \frac{\partial f_0}{\partial x}, \qquad (A\ 4.29)$$

we obtain

$$\left(-i\omega+\frac{1}{\tau}\right)a = b - \frac{e}{M\omega_0^2\tau} \qquad (A\ 4.30)$$

$$\left(-i\omega+\frac{1}{\tau}\right)b = -\omega_0^2 a - \frac{e}{M}. \qquad (A\ 4.31)$$

Hence

$$a = -\frac{e}{M\omega_0^2}\frac{\omega_0^2\tau^2 + 1 - i\omega\tau}{\omega_0^2\tau^2 + (1-i\omega\tau)^2}$$

$$= -\frac{e}{M\omega_0^2}\frac{1}{2}\left(\frac{1-i\omega_0\tau}{1-i(\omega+\omega_0)\tau} + \frac{1+i\omega_0\tau}{1-i(\omega-\omega_0)\tau}\right). \qquad (A\ 4.32)$$

Together with A 4.27 and A 4.12 this leads to equation 13.9.

B 1. STATIC DIELECTRIC CONSTANT

The method developed in § 7 for calculating the static dielectric constant ϵ_s, leading to 7.20, 7.21, and 7.33, is perfectly general. It consists in treating a large sphere of volume V macroscopically and assuming it embedded in a continuous medium with the same dielectric constant. This method can be generalized by permitting the surrounding continuous medium to have an arbitrary static dielectric constant, say ϵ_1. This contains then the two special cases $\epsilon_1 = 1$ and $\epsilon_1 = \epsilon_s$, where the sphere is in vacuum or is embedded in its own medium. The interaction of the sphere with the surroundings can be calculated electrostatically from the dipole moment **M** of the sphere (in a certain configuration of its particles), with the help of the static dielectric constant of the surroundings. Clearly, if the sphere is large enough then this does not involve any microscopic considerations. The macroscopic equation 7.14 can then be taken as a starting-point for calculation of the mean square fluctuation $\overline{M^2}$. Evaluation of the integrals in 7.14 requires the part of the free energy $F(M)$ of the system which depends on the moment M of the sphere. This free energy is identical with the corresponding electrostatic energy (cf. 3.13) and, according to 7.16 and 7.18, can be written as a sum of an internal and an external part

$$F(M) = F_i(M) + F_e(M), \qquad (B\,1.1)$$

with $F_i(M)$ given by 7.16 and

$$F_e(M) = -\tfrac{1}{2} M R; \quad R = g M. \qquad (B\,1.2)$$

The factor g is, according to Appendix A 2.19, given by

$$g = \frac{4\pi}{3V} \frac{2(\epsilon_1 - 1)}{2\epsilon_1 + 1}, \qquad (B\,1.3)$$

so that, using B 1.1–B 1.3 and 7.16,

$$F(M) = \frac{1}{2} \frac{4\pi}{V} \frac{2\epsilon_1 + \epsilon_s}{(\epsilon_s - 1)(2\epsilon_1 + 1)} M^2. \qquad (B\,1.4)$$

Sometimes calculation of the reaction field with the help of the static dielectric constant is objected to; for it is said, since the moment M fluctuates with time, an appropriate dynamic dielectric constant should be used to calculate R. This is wrong; for the general formula (that following 7.12) from which 7.14 is obtained, contains only the free energy $F(M)$, which is a static quantity. Derivation of this formula is given in reference $T1$. The reason for the absence of dynamic elements goes back to the weighting factor $\exp(-H/kT)$ in statistical mechanics, where H is the total energy of the system which is the sum of kinetic and potential energies,

$$H = H_{\text{kin}}(\dot{X}) + H_{\text{pot}}(X). \qquad (B\,1.5)$$

Here X stands for all positional coordinates and $\dot{X}$ for the respective velocities. In calculating the average of a function of positional coordinates, say $f(X)$, terms depending on $\dot{X}$ cancel,

$$\bar{f} = \int f(X) e^{-H/kT} \, dX \, d\dot{X} \Big/ \int e^{-H/kT} \, dX \, d\dot{X}$$
$$= \int f(X) e^{-H_{\text{pot}}/kT} \, dX \Big/ \int e^{-H_{\text{pot}}/kT} \, dX, \qquad (B\,1.6)$$

which means that dynamic properties do not enter in $\bar{f}$. While this is convincing in a general way, a very simple example will be given in the section B 2, which should help to illustrate these questions.

To complete calculation of $\overline{M^2}$ we introduce B 1.4 into 7.14. The integration (making use of the footnote on page 42) yields

$$\overline{M^2} = kT \frac{3V}{4\pi} (\epsilon_s - 1) \frac{2\epsilon_1 + 1}{2\epsilon_1 + \epsilon_s}. \qquad (B\,1.7)$$

For $\epsilon_1 = \epsilon_s$ and $\epsilon_1 = 1$ this equation becomes identical with 7.20 and 7.21 respectively.

Although B 1.7 has been derived for a sphere containing very many particles, it is worth noticing that if we apply it to the case of a single rigid dipole ($M = \mu$) embedded in its own medium ($\epsilon_1 = \epsilon_s$), equation B 1.7 becomes identical with Onsager's equation for rigid dipoles (eqn.

6.36 with $n = 1$, $N_0 = 1/V$). This is not accidental, for Onsager's derivation assumes absence of short range interaction. This means (i) that any two dipoles interact as if they were point dipoles, and (ii) that they are arranged so that the substance can be considered to be isotropic. In this case the interaction of a given dipole with all the others is the same as if it were interacting with a continuous medium. For the electric field produced by such a medium at the position of a given dipole can be considered as a superposition of the dipolar contributions of small volume elements. One is tempted to improve on this by including quadripole and higher contributions, owing to a not quite complete isotropy of the medium. This, however, would be inconsistent, for such contributions have a shorter range than dipolar ones. To be consistent, it would be necessary to include other types of short-range forces as well.

Separation of high-frequency contributions

The mean fluctuation $\overline{M^2}$ which has been calculated in B 1.7, can, in principle, be analysed into mean contributions at various frequencies. Calculation of this would involve the frequency-dependent complex dielectric constant $\epsilon(\omega)$. In many substances the behaviour of $\epsilon(\omega)$ indicates well-separated contributions, which can be attributed to different modes of motion. In these cases the dielectric constant $\epsilon(\omega)$ reaches, at a certain high frequency, a value ϵ_∞ and remains nearly constant up to frequencies at which infra-red absorption starts. At still higher frequencies, the dielectric constant drops to a constant value n^2, which remains nearly constant until optical absorption starts. Whenever such a clear separation occurs, then it must be possible to separate the contributions to the fluctuations $\overline{M^2}$ into well-separated groups. A very simple case is that of ionic crystals where the dynamic properties lead to normal modes in the infra-red and ultra-violet regions respectively (§ 18). Both energy and free energy are then sums of the contributions of different normal modes and the separation of $\overline{M^2}$ follows in a trivial way. In the following† a separation of high-frequency contributions will be carried through for dipolar substances. This involves, of course, macroscopic considerations only, based on two macroscopic dielectric constants ϵ_s and ϵ_∞.

We consider first the simple case of a sphere in vacuum. Here a given macroscopic moment **M** can be considered composed of two contributions, $\mathbf{M}_l$ and $\mathbf{M}_h$, indicating low and high frequency,

$$\mathbf{M} = \mathbf{M}_l + \mathbf{M}_h. \qquad (\text{B }1.8)$$

In the special case of dipolar substances, $\mathbf{M}_h$ is principally due to elastic displacement of ions and electrons, and $\mathbf{M}_l$ is due to orientation of dipoles, including whatever displacement polarization they induce. Equation

† H. Fröhlich, *Physica*, 1956, **22**, 898; B. K. P. Scaife, *Proc. Phys. Soc.*, 1957, **70 B**, 314.

STATIC DIELECTRIC CONSTANT

B 1.8 does not represent a complete definition of $\mathbf{M}_l$ and $\mathbf{M}_h$ it only defines say $\mathbf{M}_l$ if $\mathbf{M}_h$ is defined by other means; $\mathbf{M}$ is, of course, directly measurable.

The required definition of $\mathbf{M}_h$ must involve the high-frequency dielectric constant ϵ_∞ and is best done with the help of the free energy. Assume first dipolar orientation to be completely absent in the frequency range in which $\epsilon = \epsilon_\infty$. Imagine, then, a demon to prevent dipolar orientation, while he permits elastic displacement of charges. Our substance would then have a static dielectric constant ϵ_∞, and since $F_e = 0$ (using $\epsilon_1 = 1$), the free energy, denoted as $F_{ih}(M_h)$, is given by 7.16 if ϵ_s is replaced by ϵ_∞,

$$F_{ih}(M_h) = \frac{1}{2}\frac{M_h^2}{\beta(1)}; \qquad \beta(1) = \frac{3V}{4\pi}\frac{\epsilon_\infty - 1}{\epsilon_\infty + 2}. \tag{B 1.9}$$

Imagine now, for given $\mathbf{M}_h$, the demon to orient the dipoles such that the total moment becomes $\mathbf{M}$ without changing $\mathbf{M}_h$, but remember that by definition, $\mathbf{M}_l$ includes any displacement moment induced by dipolar orientation. We shall show that this definition is identical with the assumption that the total free energy (identical with the internal one since the sphere is in vacuum, $\epsilon_1 = 1$) denoted by $F_{i,lh}(\mathbf{M}_l, \mathbf{M}_h)$ is a sum of the form

$$F_{i,lh}(\mathbf{M}_l, \mathbf{M}_h) = F_{il}(M_l) + F_{ih}(M_h), \tag{B 1.10}$$

where F_{ih} is given by B 1.9 and

$$F_{il}(M_l) = \frac{1}{2}\frac{M_l^2}{\gamma(1)}; \qquad \gamma(1) = \frac{3V}{4\pi}\left(\frac{\epsilon_s - 1}{\epsilon_s + 2} - \frac{\epsilon_\infty - 1}{\epsilon_\infty + 2}\right). \tag{B 1.11}$$

To prove this we have to show that 7.16 for $F_i(M)$ is obtained by averaging $F_{i,lh}$ appropriately over all differences $\mathbf{D} = \mathbf{M}_l - \mathbf{M}_h$ but keeping $\mathbf{M} = \mathbf{M}_l + \mathbf{M}_h$ constant, i.e.

$$e^{-F_i/kT} = \text{const.} \int e^{-F_{i,lh}/kT} \, d\mathbf{D}. \tag{B 1.12}$$

Here const. represents an irrelevant constant which is independent of the M's. Now

$$\mathbf{M}_l = \tfrac{1}{2}(\mathbf{D} + \mathbf{M}); \qquad \mathbf{M}_h = \tfrac{1}{2}(\mathbf{M} - \mathbf{D}), \tag{B 1.13}$$

i.e.

$$F_{i,lh} = \frac{1}{2}\left(\frac{M_h^2}{\beta(1)} + \frac{M_l^2}{\gamma(1)}\right) = \frac{1}{8}\left\{\left(\frac{1}{\beta(1)} + \frac{1}{\gamma(1)}\right)(D^2 + M^2) + 2\left(\frac{1}{\gamma(1)} - \frac{1}{\beta(1)}\right)\mathbf{DM}\right\}. \tag{B 1.14}$$

Integration of B 1.12 and comparison of M^2-terms then gives

$$F_i = \frac{1}{2}\frac{M^2}{\beta(1) + \gamma(1)}. \tag{B 1.15}$$

Comparing this expression with 7.16 we find

$$\beta(1) + \gamma(1) = \frac{3V}{4\pi}\frac{\epsilon_s - 1}{\epsilon_s + 2}, \tag{B 1.16}$$

which is in agreement with our definitions B 1.9 and B 1.11 of $\beta(1)$ and $\gamma(1)$, as required.

An important consequence of the splitting of $F_{i,lh}$ into two terms, proportional to M_l^2 and M_h^2 respectively, is that $\overline{\mathbf{M}_l \mathbf{M}_h} = 0$ so that

$$\overline{M^2} = \overline{M_h^2} + \overline{M_l^2} \tag{B 1.17}$$

holds in addition to B 1.8.

As a next step, we consider the sphere to be embedded in a continuous medium with dielectric constant ϵ_1. We wish to split $\mathbf{M}$ again, so that besides B 1.8, B 1.17 also holds, i.e. that the free energy F_{lh} can be written as

$$F_{lh} = F_l + F_h, \tag{B 1.18}$$

where
$$F_h = \frac{1}{2} \frac{M_l^2}{\beta(\epsilon_1)}, \qquad F_l = \frac{1}{2} \frac{M_h^2}{\gamma(\epsilon_1)}. \tag{B 1.19}$$

If $\epsilon_1 \neq 1$, the free energy F_{lh} is a sum of external and internal free energies, $F_{e,lh} + F_{i,lh}$ as in B 1.1. Also $F_{e,lh}$ is given exactly by B 1.2 and B 1.3, and is thus proportional to $M^2 = (\mathbf{M}_l + \mathbf{M}_h)^2$ which contains a term proportional to $\mathbf{M}_l \mathbf{M}_h$. In order to give F_{lh} the required form B 1.18 and B 1.19, it would be wrong to use the same definitions of $\mathbf{M}_l$ and $\mathbf{M}_h$ as in the case $\epsilon_1 = 1$. For then the internal free energies would be given by B 1.9 and B 1.11, and the resultant F_{lh} would contain a term proportional to $\mathbf{M}_l \mathbf{M}_h$. This means that the definitions of $\mathbf{M}_h$ with which we started must depend on ϵ_1, and B 1.8 must be replaced by

$$\mathbf{M} = \mathbf{M}_h(\epsilon_1) + \mathbf{M}_l(\epsilon_1), \tag{B 1.20}$$

indicating that for a given $\mathbf{M}$, both $\mathbf{M}_h$ and $\mathbf{M}_l$ depend on ϵ_1. This is to be expected, for $\mathbf{M}_l$ is supposed to include the displacement polarization which it induces. This means that in order to find $\mathbf{M}_l(\epsilon_1)$ we have to add to $\mathbf{M}_l(1)$ the displacement polarization induced by the reaction field. According to B 1.2 and B 1.3 the latter is given by $g\mathbf{M}_l(\epsilon_1)$, and from electrostatics it follows that such a field produces a moment $\beta(1)g\mathbf{M}_l(\epsilon_1)$ in the sphere (cf. B 1.9). Hence

$$\left. \begin{array}{l} \mathbf{M}_l(\epsilon_1) - \beta(1)g\mathbf{M}_l(\epsilon_1) = \mathbf{M}_l(1) \\ \mathbf{M}_l(\epsilon_1) = \dfrac{\mathbf{M}_l(1)}{1 - \beta(1)g} \end{array} \right\}. \tag{B 1.21}$$

Now, for given total $\mathbf{M}$, B 1.20 holds for any ϵ_1 including $\epsilon_1 = 1$. Hence we require

$$\mathbf{M}_h(\epsilon_1) = \mathbf{M}_h(1) - \frac{\beta(1)g}{1 - \beta(1)g} \mathbf{M}_l(1). \tag{B 1.22}$$

We now make use of the fact that the free energy is a sum of $F_{i,lh}$, the value of F for $\epsilon_1 = 1$ and of $F_{e,lh}$, which is given by B 1.2, i.e. by $-\frac{1}{2}gM^2$ or using B 1.10 and B 1.11

$$F_{lh}\{\mathbf{M}_l(\epsilon_1), \mathbf{M}_h(\epsilon_1)\} = \frac{1}{2} \left(\frac{\mathbf{M}_l^2(1)}{\gamma(1)} + \frac{\mathbf{M}_h^2(1)}{\beta(1)} - gM^2 \right). \tag{B 1.23}$$

Replacing here $\mathbf{M}_l(1)$ and $\mathbf{M}_h(1)$ by $\mathbf{M}_l(\epsilon_1)$ and $\mathbf{M}_h(\epsilon_1)$ according to B 1.21 and B 1.22 and making use of B 1.20, we find that F_{lh} reduces to a sum of

STATIC DIELECTRIC CONSTANT

two squares, as required in B 1.18 and B 1.19, with (using B 1.9, B 1.11, and B 1.3)

$$\beta(\epsilon_1) = \frac{V}{4\pi}(\epsilon_\infty - 1)\frac{2\epsilon_1+1}{2\epsilon_1+\epsilon_\infty} \tag{B 1.24}$$

and

$$\gamma(\epsilon_1) = \frac{V}{4\pi}(\epsilon_s - \epsilon_\infty)\frac{(2\epsilon_1+1)^2}{(2\epsilon_1+\epsilon_\infty)(2\epsilon_1+\epsilon_s)}. \tag{B 1.25}$$

Hence, as in 7.14,

$$\overline{M_l^2}(\epsilon_1) = \int_0^\infty dM_l \int_0^\infty dM_h\, M_l^2\, e^{-F_{lh}/kT}\, M_l^2\, M_h^2 \Big/ \int_0^\infty dM_l \int_0^\infty dM_h\, e^{-F_{lh}/kT}\, M_l^2\, M_h^2, \tag{B 1.26}$$

and similarly for $\overline{M_h^2}(\epsilon_1)$, giving

$$\overline{M_h^2}(\epsilon_1) = 3kT\beta(\epsilon_1) \quad \text{and} \quad \overline{M_l^2}(\epsilon_1) = 3kT\gamma(\epsilon_1). \tag{B 1.27}$$

More explicitly, using B 1.25, the second of these equations leads to

$$\epsilon_s - \epsilon_\infty = \frac{4\pi}{3VkT}\frac{(2\epsilon_1+\epsilon_\infty)(2\epsilon_1+\epsilon_s)}{(2\epsilon_1+1)^2}\overline{M_l^2}(\epsilon_1). \tag{B 1.28}$$

If the sphere is embedded in its own medium ($\epsilon_1 = \epsilon_s$) then B 1.28 becomes

$$\epsilon_s - \epsilon_\infty = \frac{4\pi}{3VkT}\frac{(2\epsilon_s+\epsilon_\infty)3\epsilon_s}{(2\epsilon_s+1)^2}\overline{M_l^2}(\epsilon_s). \tag{B 1.29}$$

For the special case that $\epsilon_\infty = 1$, this formula is identical with 7.20. If the sphere is in vacuum ($\epsilon_1 = 1$) then B 1.28 yields

$$\epsilon_s - \epsilon_\infty = \frac{4\pi}{3VkT}\frac{\epsilon_\infty+2}{3}\frac{\epsilon_s+2}{3}\overline{M_l^2}(1). \tag{B 1.30}$$

The above equation B 1.28, or the specialized B 1.29 and B 1.30, should be used as starting-points for the calculation of the dielectric constant ϵ_s, if ϵ_∞ is assumed to be known, e.g. from experiment. If it is intended to calculate ϵ_∞ as well, then use must be made of B 1.24 and the first equation B 1.27. Equations B 1.28–B 1.30 have been derived on the assumption that $\mathbf{M}_l$ represents the dipolar contribution to the moment. No explicit use of this has been made, however, and, in fact, these equations hold always when a (relatively) low-frequency part can be clearly separated from the high-frequency dielectric constant, i.e. wherever ϵ_∞ can be clearly defined. Normally one will use either B 1.29 or B 1.30, depending on whether, for the particular case, it is easier to calculate $\overline{M_l^2}(\epsilon_s)$ or $\overline{M_l^2}(1)$. In the case of ionic crystals, for example, $\mathbf{M}_l$ represents the infra-red contribution (cf. § 18) due to displacements of ions, including the electronic displacement which they induce. ϵ_∞ is due, then, to optical contributions, i.e. $\sqrt{\epsilon_\infty}$ is the refractive index n. In this case B 1.30 is the more useful one, for $\overline{M_l^2}(1)$ represents the fluctuation of the infra-red frequency dipole moment of a sphere in vacuum which is obtained at once from the appropriate normal mode coordinate. For (cf. 18.29 using $\mathbf{M}_l = \mathbf{P}_{ir}V$), we have

$$\mathbf{M}_l(1) = e^*N\mathbf{Q}, \tag{B 1.31}$$

which can be taken as the definition of the effective charge e^* introduced

182   APPENDIX

in § 18 and which shows that the moment is due to displacement $\mathbf{Q}$ of N particles of charge e^*. The potential energy connected with this is (cf. 18.27)

$$\tfrac{1}{2} N M_{\text{red}} \omega_s^2 Q^2 = \frac{1}{2} \frac{M_{\text{red}} \omega_s^2 M_l^2(1)}{e^{*2} N}, \qquad (\text{B }1.32)$$

where M_{red} is the reduced mass and ω_s the frequency as discussed in § 18. According to statistical mechanics, the thermal average of this energy is $\tfrac{3}{2}kT$ ($\tfrac{1}{2}kT$ per degree of freedom). Hence

$$\overline{M_l^2(1)} = \frac{3kTNe^{*2}}{M_{\text{red}} \omega_s^2}. \qquad (\text{B }1.33)$$

Inserting this into B 1.30, using $\epsilon_\infty = n^2$ and $N = N_0 V$, leads at once to Szigeti's formula (18.30) which can also be written

$$\frac{\epsilon_s - n^2}{\epsilon_s + 2} = \frac{4\pi}{3} \frac{n^2 + 2}{3} \frac{(e^* N_0)^2}{(N_0 M_{\text{red}}) \omega_s^2}. \qquad (\text{B }1.34)$$

In this form we see that this formula contains no microscopic quantities, for $e^* N_0$ and $M_{\text{red}} N_0$ are effective charge and reduced mass per unit volume respectively. In this case, then, it was useful to use the sphere in vacuum, according to B 1.30, and make use of its macroscopic dynamic properties in the absence of interaction with surroundings.

The case of dipolar substances is different, for here we wish to derive ϵ_s in terms of the properties of individual dipolar molecules. In this case we shall see that B 1.29 is more useful than B 1.30 for it automatically eliminates all long-range forces, both outside and inside the sphere. As in 7.24, we first decompose $\mathbf{M}_l(\epsilon_1)$ into contributions from individual molecules. We write $\mathbf{m}(x_j, \epsilon_1)$ instead of $\mathbf{m}(x_j)$ showing that this quantity depends on ϵ_1,

$$\mathbf{M}_l(\epsilon_1) = \sum_{j=1}^{N} \mathbf{m}(x_j, \epsilon_1). \qquad (\text{B }1.35)$$

As in 7.32, we thus obtain

$$\overline{M_l^2(\epsilon_1)} = N \overline{\mathbf{m}(\epsilon_1)\mathbf{m}^*(\epsilon_1)}. \qquad (\text{B }1.36)$$

It will be remembered that $\mathbf{M}_l(\epsilon_1)$ is the moment of the sphere (embedded in ϵ_1) for a given dipolar configuration of the molecules, including the displacement polarization induced thereby in the sphere. Hence $\mathbf{m}(x_j, \epsilon_1)$ is the moment due to a particular configuration of the jth molecule, including the displacement polarization induced by it in the sphere. For this purpose the latter can be treated as a continuum of dielectric constant ϵ_∞, for ϵ_∞ refers to displacement polarization only. For a given configuration, $\mathbf{m}(x_j, \epsilon_1)$ is independent of the position of the molecule inside the sphere (apart from molecules near the surface) and is thus denoted as $\mathbf{m}(\epsilon_1)$.

Now consider the case when $\epsilon_1 = \epsilon_s$ and the molecules are spherical. By definition, then,

$$\mathbf{m}(\epsilon_s) = \boldsymbol{\mu}_i \qquad (\text{B }1.37)$$

is the inner moment of the molecule in the particular configuration. For since the displacement polarizability of a single molecule is then the same as that of the sphere (treating ϵ_∞ on a macroscopic basis) it follows from electrostatics, A 2.ii, that $\mathbf{m}(\epsilon_s)$ is independent of the size of the sphere and coincides with μ_i, as defined in § 6.

In contrast to $\mathbf{m}(\epsilon_s)$, $\mathbf{m}^*(\epsilon_s)$ is the moment of the large sphere (embedded in $\epsilon_1 = \epsilon_s$) if it is in complete equilibrium, polarized by one of its molecules kept in a state with moment $\mathbf{m}(\epsilon_s)$, i.e. if, besides the polarization through long-range forces which have already been taken into account in $\mathbf{m}(\epsilon_s)$, short-range forces too are considered. This moment will be denoted by
$$\mathbf{m}^*(\epsilon_s) = \mu_i^*; \tag{B 1.38}$$
in the absence of short range forces $\mu_i^* = \mu_i$. Equation B 1.29, with B 1.36, B 1.37, and B 1.38, now becomes ($N_0 = N/V$)
$$\epsilon_s - \epsilon_\infty = \frac{3\epsilon_s(2\epsilon_s+\epsilon_\infty)}{(2\epsilon_s+1)^2} \frac{4\pi N_0}{3kT} \overline{\mu_i \mu_i^*}. \tag{B 1.39}$$
The bar indicates averaging over all molecular configurations.

In the case of the very simple molecular model used in § 8 and in the appendix (A 2.ii) where the molecule is a sphere with dielectric constant $\epsilon_\infty = n^2$, containing a rigid point dipole† μ at its centre, μ_i is obtained in terms of the vacuum moment μ_v by equation A 2.33 (replacing there ϵ_1 by ϵ_s and ϵ_2 by ϵ_∞), i.e.
$$\mu_i = \frac{\epsilon_\infty+2}{3} \frac{2\epsilon_s+1}{2\epsilon_s+\epsilon_\infty} \mu_v. \tag{B 1.40}$$
We can also define a μ_v^* by replacing μ_i by μ_i^*, and μ_v by μ_v^* in B 1.40. Then B 1.39 becomes
$$\epsilon_s - \epsilon_\infty = \left(\frac{\epsilon_\infty+2}{3}\right)^2 \frac{3\epsilon_s}{2\epsilon_s+\epsilon_\infty} \frac{4\pi N_0}{3kT} \overline{\mu_v \mu_v^*}, \tag{B 1.41}$$
which by 8.1, and $n^2 = \epsilon_\infty$, becomes identical with 8.5. It should be mentioned that B 1.39 is more general than B 1.41 because it refers to a more general molecular model.

It will be remembered that μ_i is defined in terms of the moment within a molecular volume; this is the main advantage achieved in using $\epsilon_1 = \epsilon_s$, i.e. starting from B 1.29. In the absence of long-range correlations, μ_i^* too can be defined in terms of very few molecules. It is of advantage, therefore, to start with B 1.29 whenever long-range correlations (due to short-range forces) do not exist. In the presence of long-range correlations, e.g. when the dipoles are ordered, elimination of the long-range forces is not of advantage because then the orientation of a selected dipole $\mathbf{m}_j$ affects others over large distances, so that the moment $\mathbf{m}^*$ is not contained in a small region around the selected dipole.

B 2. REACTION FIELDS: A SIMPLE EXAMPLE

The calculation of the mean fluctuation $\overline{M^2}$ and hence of the static dielectric constant ϵ_s as outlined in B 1 and § 7, makes use of the long range interaction of a sufficiently large sphere—treated microscopically—with

† This means μ is spread over a sphere whose radius tends towards zero.

its surroundings. The latter are treated as a continuous medium (static dielectric constant ϵ_1) and the interaction in question is obtained with the help of the reaction field **R**. It has been pointed out in B 1 that the static (not the dynamic) dielectric constant ϵ_1 has to be used for this purpose, although the particles (inside the sphere) which induce the reaction field are in constant motion. Although this follows from general considerations, it is instructive to demonstrate it on a very simple model which can be treated exactly.

Consider three charges e which are bound elastically to equilibrium positions. A compensation charge $-e$ is rigidly attached to each equilibrium position. The latter are placed in a straight line, the central one being at equal distance a from the outer ones. All displacements of the three particles are in the same plane and are furthermore restricted to directions perpendicular to the line containing the equilibrium positions. The displacement of the central particle is denoted by y, that of the outer ones by x_1 and x_2 respectively. The magnitude of all displacements is assumed to be small compared with a, so that the dipolar field only has to be considered in calculating the interaction between the displaced particles.

Since the field component E of a dipole μ in the μ-direction, at a distance r in a direction perpendicular to μ, is

$$E = -\mu/r^3, \tag{B 2.1}$$

it follows that the energy of interaction between the two outer dipoles is $e^2 x_1 x_2/(2a)^3$. Their total energy $H_{1,2}$ is thus ($-fx$ is a restoring force; m_0 is the mass of a particle)

$$H_{1,2} = \tfrac{1}{2}m_0(\dot{x}_1^2+\dot{x}_2^2) + \tfrac{1}{2}f(x_1^2+x_2^2) + \frac{e^2 x_1 x_2}{8a^3}. \tag{B 2.2}$$

Introducing new variables χ, χ' by

$$\sqrt{2}\chi = x_1+x_2, \qquad \sqrt{2}\chi' = x_1-x_2, \tag{B 2.3}$$

one finds

$$H_{1,2} = \tfrac{1}{2}m_0(\dot{\chi}^2+\Omega^2\chi^2+\dot{\chi}'^2+\Omega'^2\chi'^2), \tag{B 2.4}$$

where

$$\Omega^2 = \frac{1}{m_0}\left(f+\frac{e^2}{8a^3}\right), \qquad \Omega'^2 = \frac{1}{m_0}\left(f-\frac{e^2}{8a^3}\right). \tag{B 2.5}$$

It will be noticed that only symmetrical displacements χ (not the antisymmetrical χ') interact with the central dipole, the energy of interaction being (using B 2.1)

$$I = \frac{e^2(x_1+x_2)y}{a^3} = \sqrt{2}\frac{e^2 \chi y}{a^3}. \tag{B 2.6}$$

Thus the total energy of the whole system can be written as (m is the mass of the central particle)

$$H = H_{1,2} + \tfrac{1}{2}m(\dot{y}^2+\omega^2 y^2) + \sqrt{2}\frac{e^2 \chi y}{a^3}, \tag{B 2.7}$$

where ω is the frequency of the central dipole if it does not interact with the outer ones.

REACTION FIELDS: A SIMPLE EXAMPLE

According to the principles of statistical mechanics the mean fluctuation $\overline{y^2}$ of the displacement of the central dipole is

$$\overline{y^2} = \int y^2 e^{-H/kT} d\tau \Big/ \int e^{-H/kT} d\tau, \tag{B 2.8}$$

provided the system is in thermal equilibrium. Here

$$d\tau = d\chi\, d\chi'\, dy\, d\dot\chi\, d\dot\chi'\, d\dot y. \tag{B 2.9}$$

Clearly it follows that

$$\overline{y^2} = \int_{-\infty}^{\infty} d\chi \int_{-\infty}^{\infty} dy\, y^2 e^{-H_p/kT} \frac{1}{L} = -\frac{2kT}{\omega^2}\frac{\partial}{\partial m}\log L, \tag{B 2.10}$$

where H_p is the potential energy

$$H_p = \tfrac{1}{2}m_0\Omega^2\chi^2 + \tfrac{1}{2}m\omega^2 y^2 + \sqrt{2}\,\frac{e^2 y\chi}{a^3}$$

$$= \tfrac{1}{2}m_0\Omega^2\Big(\chi + \frac{\sqrt{2}\,e^2 y}{a^3 m_0\Omega^2}\Big)^2 + \Big(\tfrac{1}{2}m\omega^2 - \frac{e^4}{m_0\Omega^2 a^6}\Big)y^2, \tag{B 2.11}$$

and

$$L = \iint_{-\infty}^{\infty} e^{-H_p/kT} d\chi dy = C\int_{-\infty}^{\infty} \exp\Big\{\frac{1}{2kT}\Big(\frac{2e^4}{m_0\Omega^2 a^6} - m\omega^2\Big)y^2\Big\} dy$$

$$= C'\Big/\Big(m\omega^2 - \frac{2e^4}{m_0\Omega^2 a^6}\Big)^{\frac{1}{2}}. \tag{B 2.12}$$

Here C and C' are independent of m. From B 2.10 and B 2.12 we find

$$\overline{y^2} = kT\Big/\Big(m\omega^2 - \frac{e^2\alpha}{a^6}\Big) \tag{B 2.13}$$

if we define a polarizability α of the outer dipoles by

$$\alpha = 2e^2/m_0\Omega^2. \tag{B 2.14}$$

This quantity α has the property that in an external homogeneous field E_e (in the direction of the displacements) the sum P of the polarization of the two outer dipoles, in the absence of interaction with the central dipole, is given by

$$P = \alpha E_e, \quad \text{with} \quad P = e(x_1 + x_2) = \sqrt{2}\,e\chi. \tag{B 2.15}$$

For such a field acts equally on both dipoles. The potential energy of interaction is thus $-\sqrt{2}\,e\chi E_e$ and the total potential energy due to a displacement χ is, using P from B 2.15 and α from B 2.14,

$$V = \tfrac{1}{2}m_0\Omega^2\chi^2 - \sqrt{2}\,e\chi E_e = \frac{1}{2}\frac{P^2}{\alpha} - PE_e. \tag{B 2.16}$$

In equilibrium $\partial V/\partial P = 0$, and B 2.15 follows.

The mean fluctuation $\overline{y^2}$ of the central dipole has been derived in B 2.13 by exact application of statistical mechanics. This expression can also be derived with the help of the method of the reaction field, by treating the outside dipoles as medium with such a static dielectric constant that the static polarizability, defined by $P = \alpha E_e$ if E_e is a static field, is given by α B 2.14. The central dipole will be treated as a particle interacting with the outside medium. A displacement y of the central particle, according to B 2.1, produces a field $E = -ey/a^3$ at the site of the outside particles, i.e. in the 'medium'. With B 2.15, therefore, the polarization calculated on the basis of the static polarizability is

$$P = -\alpha ey/a^3. \qquad (B\,2.17)$$

One half of this polarization is induced on each outer dipole. According to B 2.1 then, the reaction field R, i.e. the field at the position of the central dipole due to the polarization P of the 'medium', is, using B 2.17,

$$R = -P/a^3 = \alpha ey/a^6. \qquad (B\,2.18)$$

To obtain $\overline{y^2}$ now, according to the method of § 7 or B 1, we make use of 7.14 so that

$$\overline{y^2} = \int_{-\infty}^{\infty} y^2 e^{-F(y^2)/kT}\,dy \bigg/ \int_{-\infty}^{\infty} e^{-F(y^2)/kT}\,dy, \qquad (B\,2.19)$$

where $F(y^2)$ is the term in the free energy depending on y. As in B 1.1, it can be split into an internal and an external one,

$$F = F_i + F_e, \qquad (B\,2.20)$$

where, following B 1.2, and using B 2.18,

$$F_e = -\tfrac{1}{2}MR = -\tfrac{1}{2}eyR = -\tfrac{1}{2}\alpha e^2 y^2/a^6. \qquad (B\,2.21)$$

It will be noticed that the moment M is now given by ey. The inner free energy is simply the potential energy of the particle due to the elastic binding to its equilibrium position, i.e. according to the definition of ω

$$F_i = \tfrac{1}{2}m\omega^2 y^2. \qquad (B\,2.22)$$

Hence, from the above equations B 2.20–B 2.22,

$$F = \tfrac{1}{2}\left(m\omega^2 - \frac{\alpha e^2}{a^6}\right)y^2. \qquad (B\,2.23)$$

Inserting this into B 2.19 leads at once to the result B 2.13, obtained by direct application of statistical mechanics.

It is now easy to see that replacement of the static polarizability α by the dynamic α_d would be wrong. To derive the dynamic polarizability α_d, consider the equations of motion for P and for y. We notice that the total restoring force for a displacement χ is given by $-\partial H/\partial\chi$, using H from B 2.7, so that with the definition B 2.15 of P one obtains, using B 2.14,

$$\ddot{P} + \Omega^2 P = -\alpha\Omega^2 ey/a^3. \qquad (B\,2.24)$$

If y is periodic, say $y = y_0 e^{i\nu t}$, then the induced polarization is

$$P = -\frac{\alpha\Omega^2}{\Omega^2 - \nu^2}\frac{ey}{a^3} = -\alpha_d \frac{ey}{a^3}. \qquad (B\,2.25)$$

Since $-ey/a^3$ is the field in the 'medium' due to y, extension of the definition B 2.15 to time-dependent fields gives

$$\alpha_d = \alpha\frac{\Omega^2}{\Omega^2 - \nu^2}. \qquad (B\,2.26)$$

The frequency ν can be found by supplementing B 2.24 with the corresponding equation of motion for y. Clearly replacement of α by α_d in B 2.23 would give a result which differs from that B 2.13 obtained by the exact statistical method, and which is, therefore, wrong. While B 2.25 would normally be used as a solution in an external field, it must be noted that it does not represent the most general solution of B 2.24 because this contains also terms with period Ω, solving the homogeneous equation. In thermal equilibrium, the oscillator y must be assumed to change its phase and amplitude at irregular intervals, owing to collisions with an outside agency ('heat bath') which keep it in thermal equilibrium.

The solutions of the homogeneous equation—with appropriate factors—must be added to the solution B 2.25 to account for definite values of phase and amplitude of P after a collision. A detailed investigation of this behaviour can be used to derive the frequency spectrum of the displacement fluctuations. Calculation of average fluctuations, like $\overline{y^2}$, do not require these details, however, because they are properties of the thermal equilibrium.

B 3. DIELECTRIC LOSS

The simplest type of dielectric loss described by the Debye equations (10.15–17) is equivalent to the assumption that, in a constant field, the polarization of a medium approaches its equilibrium value exponentially with time (§ 10). Equation 13.2 for resonance absorption is obtained by assuming that equilibrium is approached through exponentially damped oscillations. Models leading to Debye loss were described in § 11. None of them gives an exact dynamic treatment in terms of a molecular model. Instead, use is made of assumptions which are equivalent to assuming exponential approach to equilibrium, but which go beyond this purely phenomenological hypothesis by providing values of the Debye relaxation time τ in terms of other physical quantities. Thus the first model of § 11 assumes the substance to contain an assembly of charged particles, each of which has two equilibrium positions separated by a potential wall of height H. These particles are in interaction with the surroundings which may be described in terms of collisions of a frequency $1/\tau_0$. If $H \gg kT$, then a particle spends an average time $\tau \gg \tau_0$ in one equilibrium position before jumping to the other (cf. 11.1). It was pointed out in 11.2 that the Debye equations must be invalid at frequencies above a limiting frequency ω_L where

$$\omega_L \sim 1/\tau_0. \tag{B 3.1}$$

Possibly this breakdown may already, make itself felt at lower frequencies.

The restriction of the validity of the Debye equations to frequencies below a certain limit ω_L is quite a general feature of all models; for an exponential approach to equilibrium necessarily involves averaging over many collisions, whatever model we use. Formulae for resonance absorption, derived in § 13, will also be limited to certain ranges of the parameters involved. Details of the limitations of the formulae for Debye or resonance absorption, and derivation of equations which hold outside the valid range, must depend on the particular model used. Investigations of this type have recently been carried out by R. A. Sack and by E. P. Gross.† These authors in particular deal with non-polar gases at temperature T, containing a small fraction of dipolar molecules with moment of inertia I. Interaction between these dipolar molecules can then be neglected. For not too high frequencies, this model leads to the Debye equations with relaxation time

$$\tau = \frac{1}{\tau_1} \frac{I}{kT}, \qquad (B\ 3.2)$$

where τ_1 is of the order of the time between two collisions of a dipole with non-polar molecules. The rotational energy of a dipole is of the order $I\omega_R^2$ if ω_R is its rotational frequency. In thermal equilibrium, its average value $\overline{I\omega_R^2}$ must be of order kT, so that

$$\overline{\omega_R^2} \simeq kT/I. \qquad (B\ 3.3)$$

The Debye equations, for this model, are certainly limited to frequencies smaller than $1/\tau_1$ but one should also expect that they are restricted to frequencies below $\sqrt{\overline{\omega_R^2}}$. Since nearly free dipolar rotation is possible only if $1/\tau_1 < \omega_R$, it should be anticipated that the limit of the Debye equations for this particular model is given by

$$\omega_L^2 \simeq \overline{\omega_R^2} \simeq I/kT. \qquad (B\ 3.4)$$

This, indeed, has been found in the above-mentioned papers.

REFERENCES‡

‡ No attempt has been made to present a complete list of references.

A1. ABRAHAM, M., and BECKER, R. *Electricity and Magnetism* (Blackie).
B1. BAUER, E., and MASSIGNON, D. *Trans. Far. Soc.* 1946, **42A**, 12.
B2. BERNAL, J. D., and FOWLER, R. H. *J. Chem. Phys.* 1933, **1**, 515.
B3. BLEANEY, B., and PENROSE, R. P. *Proc. Phys. Soc.* 1947, **59**, 418.
B4. BORN, M., and GÖPPERT-MAYER, M. *Handb. d. Phys.*, 2nd ed., **24.2**, 623 (Springer, 1933).
C1. CLAUSIUS, R. *Die mechanische Wärmelehre*, **2**, 62–97 (Vieweg, 1879).
C2. CLEETON, C. E., and WILLIAMS, N. H. *Phys. Rev.* 1934, **45**, 234.
C3. COLLIE, C. H., HASTED, J. B., and RITSON, D. M. *Proc. Phys. Soc.* 1948, **60**, 145.
D1. DANFORTH, W. E. *Phys. Rev.* 1931, **38**, 1224.
D2. DEBYE, P. *Polar Molecules* (New York, 1929).
D3. —— *Phys. Zs.* 1935, **36**, 100, 193.
E1. EYRING, H. *J. Chem. Phys.* 1935, **3**, 107.
E2. —— Ibid. 1936, **4**, 283.
F1. FRANK, F. C. *Proc. Roy. Soc.* 1935, A, **152**, 171.

† R. A. Sack, *Kolloid Z.* 1953, **134**, 16, 83; *Proc. Phys. Soc.* 1957, **70B**, 402, 414; E. P. Gross, *Phys. Rev.* 1955, **97**, 395; *J. Chem. Phys.* 1955, **23**, 1415.

REFERENCES

- F2. FRANK, F. C. *Trans. Far. Soc.* 1936, **32**, 1634.
- F3. —— Ibid. 1946, **42A**, 24.
- F4. —— and SUTTON, L. E. Ibid. 1937, **33**, 1307.
- F5. FRÖHLICH, H. *Proc. Phys. Soc.* 1942, **54**, 422; *E.R.A. Report* L/T 121, 1941.
- F6. —— *Trans. Far. Soc.* 1944, **40**, 498; *E.R.A. Report* L/T 156, 1945.
- F7. —— *J.I.E.E.* 1944, **91**, part I, 456.
- F8. —— *Proc. Roy. Soc.* 1946, A, **185**, 399; *E.R.A. Report* L/T 147, 1944.
- F9. —— *Nature*, 1946, **157**, 478; *E.R.A. Reports* L/T 157, 1945, L/T 163, 1946.
- F10. —— *Trans. Far. Soc.* 1948, **44**, 238.
- F11. —— and MOTT, N. F. *Proc. Roy. Soc.* 1939, A, **171**, 496.
- F12. —— and SACK, R. Ibid. 1944, A, **182**, 388.
- G1. GARTON, C. G. *Trans. Far. Soc.* 1946, **42A**, 56.
- G2. GEVERS, M., and DU PRÉ, F. K. Ibid. 47.
- G3. GIRARD, P., and ABADIE, P. Ibid. 40.
- G4. GROSS, B. *Phys. Rev.* 1941, **59**, 748.
- G5. GROSS, P., and HALPERN, O. *Phys. Zs.* 1925, **26**, 403.
- H1. HARTSHORN, L., MEGSON, N. J. L., and RUSHTON, E. *Proc. Phys. Soc.* 1940, **52**, 796.
- H2. HIGASI, K. *Sci. Pap. I.P.C.R.* 1936, **28**, 284.
- H3. HUBY, R. *E.R.A. Report* L/T 179, 1947.
- J1. JACKSON, W. *Proc. Roy. Soc.* 1935, A, **150**, 197.
- J2. —— Ibid. 1935, A, **153**, 158.
- J3. —— and POWLES, J. G. *Trans. Far. Soc.* 1946, **42A**, 101.
- K1. KAUZMANN, W. *Rev. mod. Phys.* 1932, **14**, 12.
- K2. KELLERMANN, E. W. *Phil. Trans. Roy. Soc.* 1940, A, **238**, 513.
- K3. KIRKWOOD, J. G. *J. Chem. Phys.* 1936, **4**, 592.
- K4. —— Ibid. 1939, **7**, 911.
- K5. —— Ibid. 1940, **8**, 205.
- K6. —— *Trans. Far. Soc.* 1946, **42A**, 7.
- L1. LEE, E., SUTHERLAND, G. B. B. M., and CHANG-KAI WU. *Proc. Roy. Soc.* 1940, A, **176**, 493.
- L2. LE FÈVRE, R. J. W. *Dipole Moments* (Methuen, 1948).
- L3. LORENTZ, H. A. *The Theory of Electrons*, § 117 (Teubner, 1909).
- L4. LYDDANE, R. H., HERZFELD, K. F., and SACHS, R. G. *Phys. Rev.* 1940, **58**, 1008.
- L5. ——, SACHS, R. G., and TELLER, E. *Phys. Rev.* 1941, **59**, 673.
- M1. MAGAT, M. *Trans. Far. Soc.* 1946, **42A**, 77.
- M2. MICHELS, A., and HAMERS, J. *Physica*, 1937, **4**, 995.
- M3. —— and KLEEREKOPER, L. *Physica*, 1939, **6**, 586.
- M4. MORGAN, J., and WARREN, B. E. *J. Chem. Phys.* 1938, **6**, 666.
- M5. MOSSOTTI, O. F. *Mem. di math. e fisica di Modena*, 1850, **24**, 2, 49.
- M6. MULLER, A. *Proc. Roy. Soc.* 1928, A, **120**, 437.
- M7. —— Ibid. 1936, A, **158**, 403.
- M8. MULLER, *Proc. Roy. Soc.* 1940, A, **174**, 137.

REFERENCES

M9. MÜLLER, F. H. *Ergebn. d. exakt. Naturw.* 1938, **17**, 164.
N1. NIX, F. C., and SHOCKLEY, W. *Rev. mod. Phys.* 1938, **10**, 1.
O1. ONSAGER, L. *J. Amer. Chem. Soc.* 1936, **58**, 1486.
O2. OSTER, G., and KIRKWOOD, J. G. *J. Chem. Phys.* 1943, **11**, 175.
P1. PAULING, L. *Phys. Rev.* 1930, **36**, 430.
P2. PELMORE, D. R. *Proc. Roy. Soc.* 1939, A, **172**, 502.
P3. PELZER, H. *Trans. Far. Soc.* 1946, **42A**, 164.
P4. —— and WIGNER, E. *Z. phys. Chem.* 1932, B, **15**, 445.
S1. SÄNGER, R. *Phys. Zs.* 1926, **27**, 556.
S2. SAXTON, J. A. and LANE, J. A. Report on 'Meteorological Factors in Radio-wave Propagation'. *Phys. Soc.* 1946, 278.
S3. SCHALLAMACH, A. *Trans. Far. Soc.* 1946, **42A**, 180.
S4. —— *Nature*, 1946, **158**, 619.
S5. SCOTT, A. H., MCPHERSON, A. T., and CURTIS, H. L. *J. Res. Bureau Stand., Wash.* 1933, **11**, 173.
S6. SILLARS, R. W. *Proc. Roy. Soc.* 1938, A, **169**, 66.
S7. SMYTH, C. P. *Trans. Far. Soc.* 1946, **42A**, 175.
S8. —— *Dielectric Constant and Chemical Structure* (New York, 1931).
S9. —— and HITCHCOCK, C. S. *J. Amer. Chem. Soc.* 1934, **56**, 1084.
S10. SUTTON, L. E. *Trans. Far. Soc.* 1946, **42A**, 170.
S11. SZIGETI, B. *E.R.A. Report.* L/E T105, 1947.
S12. —— Ibid. L/T 172, 1947.
S13. —— *Trans. Far. Soc.* 1949, **45**, 155.
T1. TOLMAN, R. C. *Statistical Mechanics*, § 141 d (Oxford, 1938).
T2. TURGEWICH, A., and SMYTH, C. P. *J. Amer. Chem. Soc.* 1940, **62**, 2468.
U1. UBBELOHDE, A. R. *Trans. Far. Soc.* 1938, **34**, 282.
V1. VAN VLECK, J. H. *Theory of Electric and Magnetic Susceptibilities* (Oxford, 1932).
V2. —— *J. Chem. Phys.* 1937, **5**, 320.
V3. —— Ibid. p. 556.
V4. —— *Phys. Rev.* 1947, **71**, 413, 425.
V5. —— and WEISSKOPF, V. *Rev. mod. Phys.* 1945, **17**, 227.
W1. WEIGLE, J. *Helv. Phys. Acta*, 1933, **6**, 68.
W2. WHIFFEN, D. H., and THOMPSON, H. W. *Trans. Far. Soc.* 1946, **42A**, 114, 122, and 166.
W3. WHITE, H., BIGGS, B. S., and MORGAN, S. O. *J. Amer. Chem. Soc.* 1940, **62**, 16.
W4. —— and MORGAN, S. O. *J. Chem. Phys.* 1937, **5**, 655.
W5. WHITEHEAD, S. *Trans. Far. Soc.* 1946, **42A**, 66.
W6. —— and HACKETT, W. *Proc. Phys. Soc.* 1939, **51**, 173.

INDEX

α-Bromo naphthalene, 123–4.
Absorption, 70 ff., 90 ff.
— coefficient, 98 ff., 117 ff., 173 ff.
Alkali halides, 158
Ammonia, 118–19.
Anisotropy of polarization, 112.
Argon, 111.
Atomic polarization, 106 ff.
Atoms, 104 ff.

Benzene, 108.
Benzophenone solution, 120–1.
Benzyl alcohol resin, 135.

Caesium halide, 158.
Calcium oxide, 158.
Camphor, 123, 124.
Carbon dioxide, 106, 107, 108, 115.
— monoxide, 105.
— tetrachloride, 108, 116, 117.
Cavity field, 25, 34, 39.
Cetyl palmitate, solution in paraffin wax, 138.
Chlorinated diphenyl, 135.
Chlorine, 105.
Chloroform, 116, 123, 124.
Clausius-Mossotti formula, 26, 28, 35, 110, 169 ff.
Copper halide, 158.

Debye equations, 70 ff., 78 ff., 120 ff., 187.
Decay function, 6 ff.
Diamond, 108, 109, 110.
Dichlor methane, 116.
Dichloral propane, 138.
Dielectric constant, 1 ff. *passim*.
— — (static), 2, 15 ff., 176 ff.
— — — general statistical theory, 36 ff., 174 ff.
— — — of dilute solutions of polar molecules in non-polar substances, 30 ff., 118 ff.
— — — of gases, 28 ff., 115 ff.
— — — of ionic crystals, 149 ff., 178, 181, 182.
— — — of mixtures, 47 ff.
— — — of polar liquids, 49 ff., 130 ff., 182 ff.
— — — of solids, 53 ff., 130 ff.
— — — sphericalmolecules, 31, 33 ff., 182 ff.
— — — temperature-dependence of, 12–13, 28 ff., 46, 48–61, 115–17, 186 ff.
— — (complex), 6, 62 ff.

Dielectric constant (complex), frequency dependence of, 8, 62 ff., 73, 115 ff., 118 ff., 130 ff., 178, 181, 187 ff.
— — — temperature-dependence of, 76 ff.
— — — relation between real and imaginary parts, 8, 162.
Di-isopropyl ketone, 134.
Dilute solutions of polar molecules, in benzene, 120.
— — — — in non-polar liquids, 30 ff., 89, 118 ff.
— — — — in paraffin, 120.
— — — — in solids, 118 ff., 125 ff.
Dipolar gases, 28 ff., 115 ff.
— interaction, 21 ff., 33 ff., 38 ff., 51 ff., 178.
— liquids, 49 ff., 83 ff., 130 ff.
— solids, 53 ff., 79 ff., 130 ff.
Dipole, 19 *passim*.
— molecular, 26, 105 *passim*.
Distribution of relaxation times, 91 ff.

Electric dipole moment, 4, 15 ff., 27 ff., 105 ff., 166 ff. *passim*.
— displacement, 3.
Energy, static electric field, 9 ff., 160.
— loss in periodic electric field, 5 ff., 13, 161.
Entropy, 9 ff.
Ethylene cyanide (solid), 133.

Fluctuations, 174, 177, 185, 187.
Forces between dipoles, 21 ff., 33, 36 ff., 48 ff.
Free energy, 9 ff., 40, 176, 177, 179, 180, 184.
— — external, 180, 186.
— — internal, 180, 186.
Frequency dependence of dielectric constant, *see* Dielectric constant (complex).

Gases, 28 ff., 110 ff., 115 ff.

Halides, 158.
Harmonic oscillator, 16, 63.
Helium, 111.
Heptane, 123.
Hindered rotation, 52.
Hydrogen chloride, 105.
— peroxide, 105.
— sulphide (solid), 132.
Hydroxyl radical, 107.

Inert gases, 108, 111.

INDEX

Internal field (Lorentz), 22 ff., 163 ff.
— — (Onsager), 25 ff., 163 ff.

Ketones, 107, 126 ff., 142 ff.
Kirkwood's formula, 49 ff., 137.
Krvpton, 111.

Long-chain molecule, 113 ff., 125 ff.
Loss angle, 14, 73 ff., 101, 121 ff.

Magnesium oxide, 158.
Methane, 116, 117.
Methyl benzoate, 123, 124.
— chloride, 116.
Models, for crystalline solids, 54 ff.
— for Debye equations, 78 ff.
— for complex dielectric constant, 63 ff.
— for static dielectric constant, 16 ff.
Molecules, 105 ff. *passim*.

Neon, 111.
Nitrogen, 111.
Non-polar molecules, 26, 105 ff.

Onsager's formula, 33 ff., 49 ff., 53, 130, 171, 177, 178.
Optical constants, 14, 162.
— polarization, 105 ff., 149.
Order-disorder transitions in dipolar crystals, 53 ff., 132, 146.
Oxygen, 108, 111.

Paraffin, 108, 113 ff., 125 ff.
Penta-methyl chlor-benzene, 138.
Pentane, 114.

Phenolic resin, 135.
Polar molecules, 26 ff., 105 ff. *passim*.
Polarizability, 28, 105 ff., 185, 186.
Polarization, 2, 64.
— displacement, 180, 182.
Polarization waves, 149 ff.
Power loss of dielectrics in periodic field, 13, 73 ff., 92 ff., 98 ff.

Rate of unimolecular reactions, 81.
Reaction field, 25, 31, 34, 41, 165, 177, 183 ff.
Refractive index, 28, 163.
Relaxation time, 73 ff., 88, 91 ff., 121 ff., 128 ff., 187, 188.
Resonance absorption, 98 ff., 173 ff.
Rubber-sulphur compounds, 135.
Rubidium halide, 158.

Self-energy, 144, 145, 155, 169.
Sodium chloride, 109, 157.
Strontium oxide, 158.
Superposition principle, 6.

Thallous chloride, 158.
Temperature-dependence of dielectric constant, *see* Dielectric constant.
Tertiary butyl chloride, 138.
Titanium oxide, 109.

Viscosity, 84 ff., 89, 123.

Water, 107, 137 ff.

Xenon, 111.